T029I884

CAMBRIDGE LIBRARY COLLECTION

Books of enduring scholarly value

Mathematical Sciences

From its pre-historic roots in simple counting to the algorithms powering modern desktop computers, from the genius of Archimedes to the genius of Einstein, advances in mathematical understanding and numerical techniques have been directly responsible for creating the modern world as we know it. This series will provide a library of the most influential publications and writers on mathematics in its broadest sense. As such, it will show not only the deep roots from which modern science and technology have grown, but also the astonishing breadth of application of mathematical techniques in the humanities and social sciences, and in everyday life.

The Theory of the Imaginary in Geometry

John Leigh Smeathman Hatton (1865–1933) was a British mathematician and educator. He studied at Hertford College, Oxford and in 1892 became Director of Evening Classes at the People's Palace in East London, a pioneering educational project that developed into East London College and later became Queen Mary College. The College awarded University of London degrees from 1902, and Hatton served as its Principal between 1908 and 1933. This book, published in 1920, explores the relationship between imaginary and real non-Euclidean geometry through graphical representations of imaginaries under a variety of conventions. This relationship is of importance as points with complex determining elements are present in both imaginary and real geometry. Hatton uses concepts including the use of co-ordinate methods to develop and illustrate this relationship, and concentrates on the idea that the only differences between real and imaginary points exist solely in relation to other points. This clearly written volume provides insights into the type of non-Euclidean geometry which was being researched at the time of publication.

Cambridge University Press has long been a pioneer in the reissuing of out-of-print titles from its own backlist, producing digital reprints of books that are still sought after by scholars and students but could not be reprinted economically using traditional technology. The Cambridge Library Collection extends this activity to a wider range of books which are still of importance to researchers and professionals, either for the source material they contain, or as landmarks in the history of their academic discipline.

Drawing from the world-renowned collections in the Cambridge University Library, and guided by the advice of experts in each subject area, Cambridge University Press is using state-of-the-art scanning machines in its own Printing House to capture the content of each book selected for inclusion. The files are processed to give a consistently clear, crisp image, and the books finished to the high quality standard for which the Press is recognised around the world. The latest print-on-demand technology ensures that the books will remain available indefinitely, and that orders for single or multiple copies can quickly be supplied.

The Cambridge Library Collection will bring back to life books of enduring scholarly value (including out-of-copyright works originally issued by other publishers) across a wide range of disciplines in the humanities and social sciences and in science and technology.

The Theory of the Imaginary in Geometry

Together with the Trigonometry of the Imaginary

JOHN LEIGH SMEATHMAN HATTON

CAMBRIDGE UNIVERSITY PRESS

Cambridge, New York, Melbourne, Madrid, Cape Town, Singapore,
São Paolo, Delhi, Dubai, Tokyo, Mexico City

Published in the United States of America by Cambridge University Press, New York

www.cambridge.org
Information on this title: www.cambridge.org/9781108013109

This edition first published 1920
This digitally printed version 2010

ISBN 978-1-108-01310-9 Paperback

THE THEORY

OF THE

IMAGINARY IN GEOMETRY

CAMBRIDGE UNIVERSITY PRESS

C. F. CLAY, Manager

LONDON : FETTER LANE, E.C. 4

NEW YORK : G. P. PUTNAM'S SONS

BOMBAY ⎫
CALCUTTA ⎬ MACMILLAN AND CO., LTD.
MADRAS ⎭

TORONTO : J. M. DENT AND SONS, LTD.

TOKYO : MARUZEN-KABUSHIKI-KAISHA

THE THEORY

OF THE

IMAGINARY IN GEOMETRY

TOGETHER WITH

THE TRIGONOMETRY OF THE
IMAGINARY

BY

J. L. S. HATTON, M.A.

PRINCIPAL AND PROFESSOR OF MATHEMATICS, EAST LONDON COLLEGE
(UNIVERSITY OF LONDON)

CAMBRIDGE
AT THE UNIVERSITY PRESS
1920

PREFACE

THE position of any real point in space may be determined by means of three real coordinates, and any three real quantities may be regarded as determining the position of such a point. In Geometry as in other branches of Pure Mathematics the question naturally arises, whether the quantities concerned need necessarily be real. What, it may be asked, is the nature of the Geometry in which the coordinates of any point may be complex quantities of the form $x + ix'$, $y + iy'$, $z + iz'$? Such a Geometry contains as a particular case the Geometry of real points. From it the Geometry of real points may be deduced (a) by regarding x', y', z' as zero, (b) by regarding x, y, z as zero, or (c) by considering only those points, the coordinates of which are real multiples of the same complex quantity $a + ib$. The relationship of the more generalised conception of Geometry and of space to the particular case of real Geometry is of importance, as points, whose determining elements are complex quantities, arise both in coordinate and in projective Geometry.

In this book an attempt has been made to work out and determine this relationship. Either of two methods might have been adopted. It would have been possible to lay down certain axioms and premises and to have developed a general theory therefrom. This has been done by other authors. The alternative method, which has been employed here, is to add to the axioms of real Geometry certain additional assumptions. From these, by means of the methods and principles of real Geometry, an extension of the existing ideas and conception of Geometry can be obtained. In this way the reader is able to approach the simpler and more concrete theorems in the first instance, and step by step the well-known theorems are extended and generalised. A conception of the imaginary is thus gradually built up and the relationship between the imaginary and the real is exemplified and developed. The theory as here set forth may be regarded from the analytical point of view as an exposition of the oft quoted but seldom explained " Principle of Continuity."

The fundamental definition of Imaginary points is that given by Dr Karl v. Staudt in his *Beiträge zur Geometrie der Lage*; Nuremberg, 1856 and 1860. The idea of (α, β) figures, independently evolved by the author, is due to J. V. Poncelet, who published it in his *Traité des Propriétés Projectives des Figures* in 1822. The matter contained in four or five pages of Chapter II is taken from the lectures delivered by the late Professor Esson, F.R.S., Savilian Professor of Geometry in the University of Oxford, and may be partly traced to the writings of v. Staudt. For the remainder of the book the author must take the responsibility. Inaccuracies and inconsistencies may have crept in, but long experience has taught him that these will be found to be due to his own deficiencies and not to fundamental defects in the theory. Those who approach the subject with an open mind will, it is believed, find in these pages a consistent and natural theory of the imaginary. Many problems however still require to be worked out and the subject offers a wide field for further investigations.

To Professor Whitehead, F.R.S., the author has to tender his sincerest thanks for much valuable help and assistance. To Professor G. S. Le Beau and Mr S. G. Soal of the East London College he is indebted for very valuable advice and criticism. To Mr J. B. Peace of the Cambridge University Press he desires to tender his warmest thanks for the great assistance which he has rendered in connection with this book.

J. L. S. H.

Chideock,
September, 1919.

CONTENTS

CHAPTER I

CHAPTER II

CHAPTER III

CHAPTER IV

CHAPTER V

CHAPTER VI

CHAPTER VII

NOTE

The reference numbers of articles in square brackets, such as Art. [22], refer to articles in the author's *Principles of Projective Geometry*. In this book, as in the *Principles of Projective Geometry*, the term "line" is used as an abbreviation for straight line.

CHAPTER I

1. Imaginary points on a real straight line.

Axiom I. *Every overlapping involution range determines a pair of points as its double points, each of which has a definite position on the base.*

This axiom does not imply that the position of these points can be graphically determined with reference to real points on the base.

Def. 1. *Each of these double points is termed an imaginary point and, considered as a pair of points, they are termed a pair of conjugate imaginary points.*

On reference to Art. [51] it will be seen that, if O is the centre of an overlapping involution of which A and A' are a pair of conjugate points—situated on opposite sides of O—then the double points of the involution are given as the positions of a point P which satisfies the relation

$$OP^2 = OA \cdot OA' = -K^2.$$

Hence, if P_1 and P_2 be these points,

$$OP_1 = +\sqrt{-1}\,K \quad \text{and} \quad OP_2 = -\sqrt{-1}\,K.$$

If O' be any other real point on the base such that $O'O = L$, then the positions of P_1 and P_2 relative to O' are given by the relations

$$O'P_1 = L + \sqrt{-1}\,K \quad \text{and} \quad O\,P_2 = L - \sqrt{-1}\,K.$$

The axiom therefore states that a definite position may be assigned to the points P_1 and P_2 on the base, and consequently a definite magnitude to the lengths $\sqrt{-1}\,K$ and $-\sqrt{-1}\,K$ measured along the base. Such lengths cannot be equal to any real lengths and therefore the points P_1 and P_2 cannot coincide with any real points situated on the base.

Similarly every other overlapping involution on the same base, which has its centre at O, has a pair of imaginary double points situated on the base at distances $\sqrt{-1}\,K'$ and $-\sqrt{-1}\,K$ from O, where K is different for each involution and may have any real value from 0 to ∞.

Hence it follows that, given any real straight line l and any real point O on it, there are on the straight line an infinite number of pairs of imaginary points. Such a system of points may be termed an imaginary system, base l, centre or mean point O.

A quantity of the form $\sqrt{-1}\,K$ is termed a purely imaginary quantity and a quantity of the form, $L + \sqrt{-1}\,K$, an imaginary or a complex quantity, where L and K are real.

Imaginary lengths. The position of an imaginary point P may be determined by its distance $O'O + OP$ from a real point O', where $O'O$ is a real length and OP an imaginary length A point P' is termed *the graph of P* if it is at a distance $O'O + OP'$ from O', where

$$\sqrt{-1}\,OP' = OP.$$

Hence, if P be the graph of P, the distance of P from O' is

$$O'O + \sqrt{-1}\,.\,OP'.$$

A real length $\sqrt{+1}\,K$ and an imaginary length $\sqrt{-1}\,K$ are incommensurable, and do not in themselves involve any relative magnitudes. Like two real incommensurable quantities $\sqrt{2}$ and 3, whose squares are commensurable, the squares of a real and an imaginary quantity may be commensurable.

The position of a real point on a given base can only be determined graphically when the unit is known in which its distance from a given point on the base is expressed. Similarly, the position of an imaginary point is only known when the unit in which its imaginary distance is expressed is also known. The units in the two cases may be regarded as $\sqrt{+1}$ and $\sqrt{-1}$. As there is no inherent relation as to magnitude between these units, the relative position of real and imaginary points on the base is indeterminate.

Lengths of the form $\sqrt{-1}\,K$ and $\sqrt{-1}\,K'$ which determine the positions of imaginary points may be combined like real lengths. (See Art. 3.) So long as the lengths considered are all real or all purely imaginary each system may be graphed in the same way, the quantities $\sqrt{+1}$ and $\sqrt{-1}$ being regarded as units in which the lengths are expressed, the only difference between these units lying in the fact that in one case the square on a line is regarded as positive and in the other negative. Thus in accordance with the conventions of Algebra the imaginary point at a distance $2K\sqrt{-1}$ from O is at double the distance from O of the point at a distance $K\sqrt{-1}$, and in accordance with the conventions of coordinate geometry the points at distances $+\sqrt{-1}\,K$ and $-\sqrt{-1}\,K$ from O are at equal distances from O on opposite sides of O.

If a second system of overlapping involutions, which have any other real point O' on the base for centre, is considered, a second system of imaginary points is obtained, which are determined by imaginary lengths measured from O' This is the system, base l, centre O', and a similar system exists for every real point on the base. No two imaginary points can coincide, when they are thus determined from different real base points. Otherwise a real and an imaginary length would be equal.

If a definite position has been assigned to an imaginary point, distances may be measured from such an imaginary point as origin. If O is a real point on a straight line and P a point on the straight line at a distance $\sqrt{-1}\,K$ from O, the point P may be taken as the centre of an involution. As in the previous case there are on the base an infinite number of imaginary points, centre P, and an infinite number of points real with reference to the centre P. None of the real points, centre P, can coincide with the real points centre O, but the imaginary points centre P will be a repetition of the imaginary points centre O.

Hence the following conception of a straight line is arrived at.

On any real straight line some point may be taken which may be termed a base point. An infinite number of real and an infinite number of imaginary points may be obtained by measuring real or purely imaginary distances from this base point. Any one of these real points may be taken as a new base point and an infinite number of real and an infinite number of imaginary points may be obtained by measuring real and purely imaginary distances from it. The real points so obtained are a repetition of the real points first obtained. The imaginary points form a new system of points. An imaginary point of the first system may be taken as a base point and an infinite number of points real with respect to this base point may be obtained and also an infinite number of points imaginary with respect to this base point. The real points with respect to this centre are imaginary points with respect to the base point first taken and are distinct from those obtained from the original base point. The imaginary system obtained from this second centre is a repetition of the imaginary points obtained from the first base point.

2. Conjugate imaginary points.

It follows from definition (1) that *every imaginary point has one and only one conjugate imaginary point.* Given an imaginary point, its conjugate imaginary point is the other double point of the involution of which it is a double point. The conjugate of a given imaginary point may be obtained by changing the sign of $\sqrt{-1}$ in the length or lengths by which the position of the point is determined. Hence imaginary points occur in pairs, viz. in pairs of conjugate imaginary points, and the connector of any pair is the base of the involution of which they are the double points. When an imaginary point is given, the involution of which it is a double point is completely determined, for the centre and constant of the involution are known.

This is true so long as the same points are regarded as real points, that is as long as the origin is not moved through an imaginary distance. Thus $OK + i . KP$ and $OK - i . KP$ give two positions of P which correspond to conjugate imaginary points. If however the origin be moved through a distance $i . KP$, these become OK and $OK - 2 . i . KP$, which distances give a pair of points which are not conjugate imaginary points according to the definition. In fact by moving the origin any two points whose distances from the origin are $a + i\beta$ and $a + i\beta'$ may be made into a pair of conjugate imaginary points. Hence conjugate imaginary points are such with respect to a given origin or with respect to certain given real points. This ought always to be stated, but, when there is no risk of misunderstanding, the limitation in question will be omitted.

3. Measurement of distances.

The positions of two imaginary points P' and Q' which have the same centre O and whose graphs are P and Q are given by the relations

$$OP' = \sqrt{-1}\, OP \text{ and } OQ' = \sqrt{-1}\, OQ.$$

The distance $Q'P'$ may be defined as $OP' - OQ' = \sqrt{-1}\, QP$.

Hence, since $QP + PQ = 0,\ Q'P' + P'Q' = 0.$ ....(1)

If R' be a third imaginary point with the same centre, it also follows that

$$P'Q' + Q'R' + R'P' = 0. \quad\quad ....(2)$$

The relations (1) and (2) are those on which all the theorems of Art. [7], which refer to points on a straight line, depend. Hence these theorems hold for imaginary points with a common centre.

Two imaginary points P', centre O_1, and Q', centre O_2, whose graphs are respectively P and Q may be determined with reference to any origin O by

$$OP' = OO_1 + \sqrt{-1}\, O_1P \text{ and } OQ' = OO_2 + \sqrt{-1}\, O_2Q.$$

The distance $Q'P'$ may be defined as $OP' - OQ'$, hence

$$Q'P' = OO_1 + \sqrt{-1}\, O_1P - OO_2 - \sqrt{-1}\, O_2Q$$
$$= O_2O_1 + \sqrt{-1}\,\{O_1P - O_2Q\}$$
$$= O_2O_1\{\sqrt{1} - \sqrt{-1}\} + \sqrt{-1}\, QP.$$

From this it follows that if P', Q', R' be any three points on a straight line, real or imaginary,

$$P'Q' + Q'P' = 0$$

and

$$P'Q' + Q'R' + R'P' = 0.$$

Hence in the most general case all the theorems of Art. [7] which refer to distances of points on a straight line are true for imaginary points.

If O, P, and Q are three collinear real points, the point M given by the relation $OM = \tfrac{1}{2}\{OP + OQ\}$ is defined as the middle point of PQ.

Similarly, if O be any real point and P' and Q' a pair of imaginary points, centres O_1 and O_2, whose graphs are P and Q, M' the middle point of $P'Q'$ is defined as being the point given by

$$OM' = \frac{OO_1 + OO_2}{2} + \sqrt{-1}\left\{\frac{O_1P + O_2Q}{2}\right\}.$$

This point will be shown, Art. 6, to be the harmonic conjugate of the point at infinity on the base with respect to P' and Q'.

The product of the distances of a pair of conjugate imaginary points P' and Q', mean point M, from any real point O on the line $P'Q'$ is a positive real quantity. For this product is

$$(OM + i\,.\,MP)\,(OM - i\,.\,MP) = OM^2 + MP^2.$$

4. Determination of the position of imaginary points by means of ratios.

I. Let A' be any point on a given real ine and B' a point at a purely imaginary distance $A'B'$ from A', and P' a point at a purely imaginary distance $A'P'$ from A'. Then $B'P' = A'P' - A'B'$.

The ratio of P' with reference to A' and B' is $\dfrac{A'P'}{B'P'}$ or $\dfrac{A'P'}{A'P' - A'B'}$.

The ratio in this case is a real quantity.

II. Let A' be any point on a given real line, B and O points at real distances $A'B$ and $A'O$ from A', and P' an imaginary point at a distance $A'O + OP'$ from A', where OP' is imaginary.

Then
$$A'P' = A'O + OP'$$

and
$$BP' = BA' + A'O + OP'$$
$$= BO + OP',$$

$$\therefore \; \frac{A'P'}{BP'} = \frac{A'O + OP'}{BO + OP'} = \frac{A'O + OP'}{A'O + OP' - A'B}.$$

The ratio of P' with respect to A and B in this case is a complex quantity.

In either case when the ratio of a point and the positions of the reference points are given, the position of the point is uniquely determined. The ratio of the conjugate imaginary point of P' is obtained by changing the sign of the imaginary part of the ratio.

If lengths are expressed in imaginary units a pair of conjugate imaginary points, whose positions would ordinarily be determined by lengths $a + \sqrt{-1}\,a'$ and $a - \sqrt{-1}\,a'$, are determined by lengths $\sqrt{-1}\,a - a'$ and $\sqrt{-1}\,a + a'$, i.e. the distances which determine the points differ only in the sign of the real part.

5. Anharmonic ratios of real and imaginary points on a real straight line.

In Art. [11] the anharmonic ratio of four real collinear points A, B, C, D was defined as being, with origin O,

$$(ABCD) = \frac{OC - OA}{OC - OB} : \frac{OD - OA}{OD - OB} = \lambda \text{ (suppose).}$$

The range $(ABCD)$ was defined as being harmonic if λ had the value -1.

DEF. 2. *If A', B', C', D' be the graphs of four collinear points real or imaginary whose positions are determined by $OA_1 + iA_1A'$; $OB_1 + iB_1B'$; etc., their anharmonic ratio is defined as being*

$$\frac{OC_1 + iC_1C' - OA_1 - iA_1A'}{OC_1 + iC_1C' - OB_1 - iB_1B'} : \frac{OD_1 + iD_1D' - OA_1 - iA_1A'}{OD_1 + iD_1D' - OB_1 - iB_1B'} = \lambda \text{ (suppose).}$$

Under certain circumstances this anharmonic ratio may have the value -1, in which case the range is said to be harmonic.

Let C' and D' be real points, i.e. let $C_1C' = D_1D' = 0$. Let A and B be a pair of conjugate imaginary points and let O be their centre. Then $OA_1 = OB_1 = 0$, $A_1A' = OA'$, $B_1B' = OB'$ and $OA' = -OB'$. Hence

$$\frac{OC' - iOA'}{OC' + iOA'} : \frac{OD' - iOA'}{OD' + iOA'} = \lambda.$$

If $\lambda = -1$,
$$\frac{OC' - iOA'}{OC' + iOA'} + \frac{OD' - iOA'}{OD' + iOA'} = 0.$$

$$\therefore \; -OC' . OD' = OA'^2 = OB'^2.$$

Therefore if C' and D' are on different sides of O so that $-OC' . OD'$ is positive, then OA' and OB' are real and $i . OA'$ and $i . OB'$ give a pair of conjugate imaginary points, which are the double points of an overlapping involution of which O is the centre and of which C' and D' are a pair of conjugate points.

Conversely *the imaginary double points of an overlapping involution are harmonic conjugates of every real pair of conjugate points of the involution.* Hence it is seen that *every pair of conjugate imaginary points may be determined as the common harmonic conjugates of two pairs of real points.*

A given pair of real points are conjugate points of an infinite number of involutions with imaginary double points. The double points of these involutions are harmonic conjugates of the given pair of points. Hence *a given pair of real points has an infinite number of pairs of harmonic conjugates which are pairs of conjugate imaginary points.*

From the definition of the anharmonic ratio of a range of real and imaginary points it follows that *if three points of a range are given and likewise the value of the anharmonic ratio of these points with a fourth point—which anharmonic ratio may have a real, an imaginary or a complex value—the position of the fourth is uniquely determined.*

It likewise follows that *the theorems, Art.* [12], *in regard to the change of the value of an anharmonic ratio, when the order of the points is changed, are true when the range is wholly or in part imaginary.* Also if $(A'B'C'D') = (ABCD)$ and $(A'B'C'E') = (ABCE)$ it follows that $(A'B'E'D') = (ABED)$.

6. Projective ranges.

DEF. 3. *Two ranges of real or imaginary points are said to be projective when the anharmonic ratio of four points of one range is equal to the anharmonic ratio of the four corresponding points of the other range.*

If A, B, C, three points of one range are given, and three corresponding points A', B', C' of the other, it is always possible, given a fourth point P of the first range, to determine uniquely a fourth point P' of the second, such that $(ABCP) = (A'B'C'P')$. Hence *two sets of three points determine two projective ranges and to each point of one range corresponds one and only one point of the other.*

As a particular case there is in each of two projective ranges, one point termed the vanishing point, which corresponds to the point at infinity in the other range. The vanishing points may be real or imaginary.

To find the harmonic conjugate of the point at infinity on a straight line with respect to two imaginary points on the line.

Let P' and Q' be the imaginary points, centres O_1 and O_2, and let P and Q be their graphs. Take any real origin O and let X' be the required point. Then

$$\frac{OP-OX'}{OQ'-OX'} : \frac{OP'-\infty}{OQ'-\infty} = -1,$$

$$\therefore \ OP'+OQ'=2.OX',$$

$$\therefore \ OX'=\frac{OO_1+OO_2}{2}+\sqrt{-1}\ \frac{O_1P+O_2Q}{2}$$

This point was defined (Art. 3) as the middle point of $P'Q'$.

If a real point O, the point at infinity and a pair of conjugate imaginary points A and A' form a harmonic range, $OA = -OA'$, and therefore O must be the mean point of A and A'.

Any two conjugate imaginary points are harmonic conjugates of their graphs.

Let P', P_1', centre O, be the two points and P and P_1 their graphs.

Then

$$(P'P_1'PP_1) = \frac{OP'-OP}{OP_1'-OP} : \frac{OP-OP_1}{OP_1'-OP_1}$$

$$= \frac{i.OP-OP}{-i.OP-OP} : \frac{i.OP+OP}{-i.OP+OP}$$

$$= \frac{i-1}{-i-1} : \frac{i+1}{-i+1}$$

$$= -1.$$

It follows that if P', P_1' be a pair of conjugate imaginary points, O their mean point, P and P_1 their graphs and ∞ the point at infinity on the line, then

$$(PP_1O\infty) \doteq (P'P_1'O\infty) = (PP_1P'P_1') = -1.$$

7. Extended conception of a real involution.

Every real involution contains two branches, one real and one purely imaginary. In one branch pairs of conjugate points overlap. In the other they do not. In the latter branch there are a pair of double points, which, according to the nature of the branch in which they occur, are real or imaginary.

If the first definition of an involution (Art. [51]) be taken and O be the point corresponding to the point at infinity, and P and P' a pair of conjugate points, then the product $OP.OP'$ is constant. There are two cases to be considered according as this constant is positive or negative.

(1) Let $\qquad\qquad OP \cdot OP' = K^2.$

(*a*) For real lengths, OP and OP', there are pairs of conjugate points which do not overlap, and a pair of real double points E and F given by $OE = - OF = K$.

This is the real branch.

(*b*) If $\qquad OP = i \cdot OP_1$ and $OP' = i \cdot OP_1',$
then $\qquad\qquad\qquad\qquad OP_1 \cdot OP_1' = - K^2.$

The points then belong to an imaginary branch of the involution, which has no double points and in which corresponding segments overlap.

(2) Let $\qquad\qquad OP \cdot OP' = - K^2.$

(*a*) For real lengths, OP and OP', there are pairs of conjugate points which overlap, and this branch of the involution has no double points.

(*b*) If $\qquad OP = i \cdot OP_1$ and $OP = i \cdot OP_1',$
then $\qquad\qquad\qquad\qquad OP_1 \cdot OP_1 = + K^2.$

The points then belong to an imaginary branch of the involution, in which there are a pair of double points whose graphs, E_1 and F_1, are given by $OE_1 = - OF_1 = K$. The corresponding segments of this branch do not overlap. Hence in an involution, the constant and centre of which are real, to a real point P corresponds a real point P and to a purely imaginary point corresponds a purely imaginary point.

The double points real or imaginary of a real involution are harmonic conjugates of every pair of conjugate points, real or purely imaginary, of the involution.

(1) If the double points and the pair of conjugates are all real this has been proved in Art. [51].

(2) If the double points are imaginary and the pair of conjugates are real the result is that of Art. 5.

(3) If the double points are imaginary and the pair of conjugates are also purely imaginary the case resolves itself into case (1) where each length is expressed in imaginary units of length.

If the second definition of an involution, Art. [51], be taken the preceding may be deduced as follows:

If two superposed projective ranges $A, B, C \ldots$ and $A', B', C' \ldots$ are such that $(ABCA') = (A'B'C'A)$, then every pair of corresponding points mutually correspond and the ranges form an involution.

The point P of the first corresponding to the point B of the second is given by

$$(AA'B\overset{.}{P}) = (A'AB'B),$$

but

$$(A'AB'B) = (AA'BB'),$$

$$(AA'BP) = (AA'BB'),$$

P coincides with B'.

Let O be the point which corresponds to the point at infinity, then

$$(AB' \infty\, O) = (A'BO \infty\,),$$

$$\therefore\quad OA \cdot OA' = OB \cdot OB' = \text{a constant}.$$

This constant may be any quantity real, imaginary, or complex in the most general case. If it is real and the point O is also real, the involution is real and the preceding results are obtained.

The condition that a system of points may form an involution may also be expressed in ratios. Let b, b', c, c', d, d' be the ratios of B, B', C, C', D, D' referred to A and A' as origin. Then since

$$(AA'BC) = (A'AB'C'),$$

$$bb' = cc' = \text{a constant}.$$

If the points A, A'; B, B'; ... are real but the involution is overlapping, the position of the double points is given by

$$X^2 = -K^2 = bb' = cc' = \dots,$$

$$\therefore\quad X_1 = \sqrt{-1}\,K \text{ and } X_2 = -\sqrt{-1}\,K.$$

Conversely if $(AA'BC) = (A'AB'C')$, then AA', BB', CC' form an involution.

Every point on a line has a conjugate in every involution on the line. If the conjugate of a point $a + ia'$ with respect to an involution, constant K, is sought, this point is found to be $\dfrac{K}{a^2 + a'^2}(a - ia')$

An imaginary involution, i.e. an involution in which the constant is a complex quantity and whose centre may be an imaginary point, has a pair of imaginary double points, which are equally distant from the centre.

If $K + iK'$ be the constant of the involution, the distances of the double points from the centre are

$$\pm \frac{1}{\sqrt{2}} \{\sqrt{\sqrt{K^2 + K'^2} + K} + i\sqrt{\sqrt{K^2 + K'^2} - K}\},$$

where the positive sign has to be given to the square roots.

8. Analytical expressions.

Every imaginary point in a real plane whose coordinates are imaginary is a double point of an involution on a real base.

Let the coordinates of the point be $a+ib$ and $c+id$. The equation of the straight line joining this point to the point whose coordinates are $a-ib$ and $c-id$ is

$$\begin{vmatrix} x & y & 1 \\ a+ib & c+id & 1 \\ a-ib & c-id & 1 \end{vmatrix} = 0$$

or

$$\begin{vmatrix} x & y & 1 \\ a & c & 1 \\ b & d & 0 \end{vmatrix} = 0.$$

This is the equation of a real line passing through the imaginary point.

The real point (a, c) is on this straight line and the distance of the point $a+ib$, $c+id$ from this point is $i\sqrt{b^2+d^2}$. Therefore the point $a+ib$, $c+id$ is a double point of the overlapping involution on the straight line

$$\begin{vmatrix} x & y & 1 \\ a & c & 1 \\ b & d & 0 \end{vmatrix} = 0,$$

whose centre is the point (a, c) and whose constant is $-(b^2+d^2)$.

The equation of the real line through the point $a+ib$, $c+id$ may also be written as

$$\frac{x-a}{b} = \frac{y-c}{d}.$$

The preceding shows that the connector of a pair of conjugate imaginary points, defined as points whose coordinates differ only in the sign of the imaginary part, is a real straight line.

Coordinates of an imaginary point.

Let the coordinates of an imaginary point P be x_1+ix_2, y_1+iy_2. Construct the real point Q (x_1, y_1). Then P may be found by drawing through Q, QN equal to ix_2 and from N, PN equal to iy_2. Since OM and MQ are real $OQ=\sqrt{x_1^2+y_1^2}$ and since QN and NP are purely imaginary $QP=i\sqrt{x_2^2+y_2^2}$.

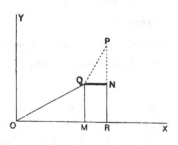

Hence P may be constructed by measuring along OQ, which makes an angle $\tan^{-1}\dfrac{y_1}{x_1}$ with the axis of x, a length $\sqrt{x_1^2+y_1^2}$, and afterwards measuring along QP, which makes an angle $\tan^{-1}\dfrac{y_2}{x_2}$ with the axis of x, an imaginary length $i\sqrt{x_2^2+y_2^2}$.

Let $\tan a=\dfrac{y_1}{x_1}$, $\tan \beta=\dfrac{y_2}{x_2}$ and $\tan \gamma=\dfrac{x_2}{y_2}$ so that $\beta+\gamma=\dfrac{\pi}{2}$

Then the coordinates of P may be written

$$x_1 + ix_2, \quad x_1 \tan a + ix_2 \tan \beta,$$

or

$$x_1 + iy_2 \tan \gamma, \quad x_1 \tan a + iy_2.$$

OQ and QP may be termed *the principal coordinates* of the point P.

Analytically the representation of all the points on a straight line (real) is as follows :

If $x_1 y_1$ and $x_2 y_2$ are any two real points which determine a straight line, the coordinates of any real point on the line may be obtained by giving all real values to λ in $\dfrac{x_1 + \lambda x_2}{1 + \lambda}, \dfrac{y_1 + \lambda y_2}{1 + \lambda}$. If λ has the value $a + ib$, where a and b may have any real values, the coordinates of all points real or imaginary on the line are of the form

$$\frac{x_1 + (a + ib)\, x_2}{1 + (a + ib)} \text{ and } \frac{y_1 + (a + ib)\, y_2}{1 + (a + ib)}.$$

If A, A' and B, B' be pairs of conjugate imaginary points and C and D a pair of real points, then the involution determined by

(1) *A, A' and B, B', as pairs of conjugate points, is real and has real double points.*

(2) *A, A' and C, D, as pairs of conjugate points, is real and has real double points.*

(3) *A, B and A', B', as pairs of conjugate points, is real.*

Let A, A', B, B', C, D, be determined by $a + ia'$, $a - ia'$, $b + ib'$, $b - ib'$, c, d.

Let the centre O of the involution be determined by x.

(1) In this case $OA \cdot OA' = OB \cdot OB'$.

Therefore $$x = \frac{a^2 + a'^2 - b^2 - b'^2}{2(a - b)}.$$

Hence the centre of the involution is real and, as the product of the distances of a pair of conjugate imaginary points from any real point on the line is real and positive, the constant of the involution is positive and the double points are real.

(2) In this case $OA \cdot OA' = OC \cdot OD$.

Therefore $$x = \frac{a^2 + a'^2 - dc}{2a - d - c}.$$

Hence the centre of the involution is real and as in case (1) the double points are real.

(3) In this case $OA \cdot OB = OA' \cdot OB'$.

Therefore $$x = \frac{ab' + ba'}{a' + b'}.$$

Hence the centre of the involution is real. The constant of the involution is $(a - x)(b - x) - a'b'$. This is real. Hence the involution is real.

From (1) it follows that *two pairs of conjugate imaginary points have always a pair of real harmonic conjugates.* (See also Art. 31 (3).)

From (2) it follows that *a pair of conjugate imaginary points and a pair of real points have always a pair of real harmonic conjugates.* (See Art. 31 (3).)

From (3) it follows that *two imaginary points have always one pair of harmonic conjugates, which are either a pair of real points or a pair of conjugate imaginary points.*

9. Imaginary straight lines.

DEF. 4. *The locus of the double points (imaginary) of the over-lapping involutions in which an overlapping involution pencil (real) is cut by real transversals is a pair of imaginary straight lines. Each part of this locus, which is continuous, constitutes an imaginary straight line.*

The two straight lines considered as a pair of straight lines are termed a pair of conjugate imaginary straight lines.

Hence it follows that an imaginary straight line is determined by an imaginary point, which is a double point of an involution, and a real point, the vertex of the involution pencil. The conjugate imaginary line is determined by the conjugate imaginary point, the other double point of the involution, and by the same real point, the vertex of the involution pencil. Hence two conjugate imaginary straight lines differ as to their determining elements only in the sign of the imaginary part. An imaginary line considered in reference to the involution pencil is spoken of as a double ray of the pencil.

From the definition it also follows that only one imaginary straight line passes through a given real and a given imaginary point.

The straight line joining the centre of an involution range to the vertex of the pencil is called the mean line of the two conjugate imaginary lines, which join its double points to the vertex of the pencil, This mean line is real. As there are an infinite number of involutions on the same base which have the same centre, there are an infinite number of pairs of imaginary straight lines through a given real point which have the same mean line.

Hence given any real point L and any real straight line o through it, there are an infinite number of pairs of imaginary straight lines which pass through L and which have o as their mean line. Such a system of straight lines may be termed an imaginary system *centre L, mean line o.*

It also follows from the definition that every imaginary straight line passes through a real point, viz. the vertex of the involution pencil of which it is a double ray, and that a pair of conjugate imaginary straight lines intersect in a real point, viz. the vertex of the pencil of which they are the double rays.

10. *To determine the straight line which joins an imaginary point to a real point.*

Construct the involution of which the imaginary point is a double point. (1) If the real point lies on this straight line the required line is the real base of the involution. (2) If the real point does not lie on the base construct an involution pencil by joining pairs of real conjugate points of the involution to the real point. The required line is one of the imaginary double rays of this involution pencil.

To determine the point of intersection of a given imaginary straight line with a given real straight line.

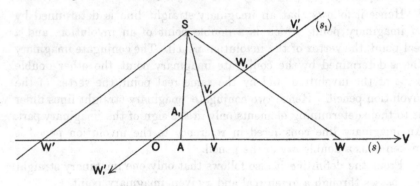

Let the given imaginary straight line be determined as the connector of one of the imaginary double points E of an involution, centre V, constant $- K^2$, situated on a straight line s, with a real point S. Let s_1 be the given real straight line. Let V' and W_1' be the points at infinity on s and s_1 respectively and let ss_1 be O.

Construct V_1 and V_1' the projections of V and V' from S on s_1. Draw SW' parallel to s_1 to meet s in W'. Take W the conjugate of W' in the involution on s. Join W to S to meet s_1 in W_1. Then W_1 is the centre of the involution into which the involution on s is projected from S on s_1. Also $W_1V_1'. W_1V_1 = - M^2$, the constant of this involution on s_1.

Hence, if E be the double point of the involution on s remote from O, the point E_1 in which SE meets s_1 is at a distance $V_1'E_1$ from V_1' such that $V_1'E_1 = V_1'W_1 - \sqrt{\overline{W_1V_1}.\overline{W_1V_1'}}$, the negative sign being given to the square root since $V_1'W_1$ is drawn in the opposite direction to VW.

The following relations hold between the lengths in the figure :

(a) $VW'. VW = - K^2$, $W_1V_1. W_1V_1' = - M^2$.

(b)   Since $(V'W'VW) = (V_1'W_1'V_1W_1)$ and V' and W_1' are at infinity,
$$\frac{VW}{W_1V_1} = \frac{VW'}{W_1V_1'} = \frac{W'W}{V_1'V_1},$$

and $\quad \dfrac{VW}{VW'} = \dfrac{W_1V_1}{W_1V_1'} = \dfrac{VW^2}{-K^2} = \dfrac{W_1V_1^2}{-M^2} = \dfrac{-K^2}{VW'^2} = \dfrac{-M^2}{W_1V_1'^2} = \lambda$ (suppose).

(c)   Since $\qquad (V'W'OV) = (V_1'W_1'OV_1),$

$$OW' \cdot OV_1' = VW' \cdot V_1V_1' = WW' \cdot W_1V_1' = AW' \cdot A_1V_1',$$

where A and A_1 are any pair of real points on s and s_1 collinear with S.

If a range composed wholly or partly of imaginary points be projected from a real point by imaginary and real straight lines upon any other real straight line so that a second range is formed on this straight line, the anharmonic ratios of any four corresponding points of these ranges are equal.

It will be proved that if, as in the preceding, V be the centre of the involution which determines an imaginary point E, and W_1 the centre of the involution which determines the corresponding point E_1, then $W'E \cdot V_1'E_1 = $ a constant, where W' and V_1' are the points on the two bases which in a projection from S correspond to the points at infinity. It will also be shown that this constant is equal to the product of the distances of any pair of real corresponding points from W' and V_1' respectively.   Hence since W' and V_1' are fixed points the theorem is true.

In the figure

$$W'E = W'V + \sqrt{VW \cdot VW'} \qquad V_1'E_1 = V_1'W_1 - \sqrt{W_1V_1 \cdot W_1V_1'}$$

$$= W'V\left(1 - \sqrt{\dfrac{VW}{VW'}}\right) \qquad = V_1'W_1\left(1 + \sqrt{\dfrac{W_1V_1}{W_1V_1'}}\right)$$

$$= W'V(1 - \sqrt{\lambda}), \qquad = V_1'W_1(1 + \sqrt{\lambda}).$$

Therefore $\qquad W'E \cdot V_1'E_1 = W'V \cdot V_1'W_1(1 - \lambda).$

But $\qquad \cdot \lambda = 1 - \dfrac{W_1V_1}{W_1V_1'} = \dfrac{V_1V_1'}{W_1V_1'}.$

Therefore $W'E \cdot V_1'E_1 = W'V \cdot V_1'V_1 = W'O \cdot V_1'O = W'A \cdot V_1'A_1.$

This theorem may be stated as follows:

If through any real point a system of real and of imaginary straight lines be drawn, these straight lines determine equianharmonic ranges on all real transversals.

It also follows that *projective ranges as defined in Art.* 6 *are also projective according to the usual conception.*

Analytical verification of Art. 10.

Let S be the vertex of the pencil and V the centre of the involution on OV. Take OV for axis of x and the given real line OV_1 for axis of y.

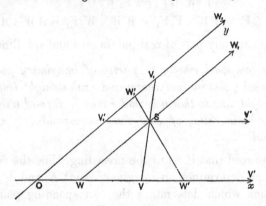

Let V' and W_1 be the points at infinity on the axes. Project these from S into V_1' and W. Let SV meet Oy in V_1.

Then V and V' are conjugate points of the involution on Ox, and therefore V_1 and V_1' are conjugate points of the involution on Oy.

Let W_1' be the centre of the involution on Oy.

Then W_1' and W_1 are conjugate points of the involution on Oy, and W and W' are conjugate points of the involution on Ox.

Since S is a given point, OW (l) and OV_1' (m) are given.

Also the constant of involution on Ox ($-b^2$) is given and the point V (a).

Hence $OW' = a + \dfrac{b^2}{a-l}$, $\therefore OW_1' = \dfrac{m(b^2 + a^2 - al)}{b^2 + (a-l)^2}$ and $OV_1 = \dfrac{ma}{a-l}$.

Therefore $V_1 W_1' = \dfrac{-mlb^2}{(a-l)\{b^2 + (a-l)^2\}}$ and $V_1' W_1' = \dfrac{ml(a-l)}{b^2 + (a-l)^2}$.

Hence $-M^2$ the constant of the involution on Oy is $-\dfrac{m^2 l^2 b^2}{\{b^2 + (a-l)^2\}^2}$.

Hence the distances of the double points of the two involutions from O are respectively $a + ib$, $a - ib$, and

$$\frac{m(b^2 + a^2 - al)}{b^2 + (a-l)^2} + i\frac{mlb}{b^2 + (a-l)^2}, \quad \frac{m(b^2 + a^2 - al)}{b^2 + (a-l)^2} - i\frac{mlb}{b^2 + (a-l)^2}.$$

But the line joining S (l, m) to ($a + ib$, 0) meets Oy in a point distant

$$\frac{m(a^2 + b^2 - al) - imbl}{(a-l)^2 + b^2}$$

from O. This shows that the geometrical construction of Art. 10 agrees with the analytical conception of an imaginary straight line.

If E and E_1 be collinear double points

$$V_1'E_1 = ml\,\frac{(a-l)-ib}{b^2+(a-l)^2}\text{ and }WE = a - l + ib.$$

$\therefore\ V_1\,E_1\,.\,WE = ml$, which is a constant for all involutions. This confirms the result of Art. 10.

11. *To determine the straight line connecting any two imaginary points.*

Construct the two involutions of which the two given imaginary points are each respectively a double point.

(1) If the two involutions have the same base this real line is the connector of the two imaginary points.

(2) If the two involutions have different bases, these involutions are in real perspective in two ways (Art. [60]) and in one of these perspectives the two given imaginary points are corresponding points. The required imaginary straight line is the connector of the centre of this perspective, which is real, with either of the two given imaginary points.

From this it follows that *only one straight line can be drawn to join two imaginary points and that it passes through a real point, viz., one of the centres of perspective of the involutions, and therefore no two imaginary straight lines can include a space.*

To determine the point of intersection of any two imaginary straight lines.

Construct the two involution pencils of which the two given imaginary straight lines are each respectively a double ray.

(1) If the two involution pencils have a common vertex, this vertex, which is real, is the point of intersection of the lines.

(2) If the two involution pencils have different vertices, these involutions are in real perspective in two ways (correlative of Art. [60]) and in one of these perspectives the two given imaginary lines are corresponding rays. The point of intersection of these lines is therefore a double point of the overlapping involution determined by the two involution pencils on one of their real axes of perspective.

Hence two imaginary straight lines intersect in only one point and this point lies on a real straight line, viz., one of the axes of perspective of the determining involutions.

12.   Summary of properties of imaginary points and straight lines.

(1) *Every imaginary point contains one and only one real straight line.*

This line is the base of the involution of which the imaginary point is a double point and it is the connector of the point and its conjugate imaginary point. There can be no other real straight line through the point for, if there were, the point would be a real point.

(2) *The connector of a pair of conjugate imaginary points is real.*

(3) *Every real straight line, that contains an imaginary point, contains its conjugate.*

(4) *An imaginary straight line meets all real straight lines in imaginary points except those which pass through its one real point.*

(5) *The connector of a pair of imaginary points is real or imaginary; if real it contains the conjugates of both the points.*

(6) *A system of real and imaginary points on a real straight line is projected from a real point, not on the line, upon another real straight line, into a system in which real points correspond to real points and imaginary points to imaginary points.*

(7) *The connector of a pair of imaginary points and the connector of their conjugate imaginary points*

Every imaginary straight line contains one and only one real point.

This point is the vertex of the involution pencil of which the imaginary line is a double ray and it is the point of intersection of the line and its conjugate imaginary line. There can be no other real point on the straight line for, if there were, the line would be a real straight line.

The point of intersection of a pair of conjugate imaginary straight lines is real.

Every real point, that contains an imaginary straight line, contains its conjugate.

The only real points, the connectors of which to an imaginary point are real, are those which lie on the one real line through the point.

The intersection of a pair of imaginary lines is real or imaginary; if real it contains the conjugates of both the lines.

A system of real and imaginary straight lines through a real point is cut by a real transversal in points which, when connected to another real point, give real straight lines corresponding to real straight lines and imaginary straight lines to imaginary straight lines.

The point of intersection of a pair of imaginary straight lines and the point of intersection of their con-

are conjugate imaginary straight lines.

(8) *The imaginary double points of an involution are harmonic conjugates of every pair of conjugate points of the involution.*

(9) *If a pair of conjugate imaginary points coincide, they coincide in a real point.*

jugate imaginary lines are conjugate imaginary points.

The imaginary double rays of an involution pencil are harmonic conjugates of every pair of conjugate rays of the involution pencil.

If a pair of conjugate imaginary lines coincide, they coincide in a real line.

13. *In two projective pencils with real vertices, which have two pairs of real corresponding rays, all pairs of corresponding rays, real or imaginary, intersect on a straight line (real) if the ray joining the vertices of the pencils is a self-corresponding ray.*

Let S and S' be the vertices of the pencils. Then by Art. [34], the real pairs of self-corresponding rays intersect on a straight line s. Let an imaginary ray e of the first pencil meet s at E and the corresponding imaginary ray of the second pencil e' meet s at E'. Let the two pairs of real corresponding rays meet s in A, B and let SS' meet s in C.

Then $(ABCE) = (ABCE')$.

Therefore by Art. 5, E and E' coincide and e and e' meet on s.

Correlatively it may be proved that *if the point of intersection of the real bases of two projective ranges, which have two pairs of real corresponding elements, is a self-corresponding point, the connectors of all pairs of corresponding points of the ranges (real or imaginary) pass through a real point.*

If two imaginary lines are corresponding rays of two projective pencils with three pairs of corresponding real rays, the two imaginary straight lines of which they are conjugate imaginary lines are also corresponding rays of the pencils.

Let e and e' be the first pair of corresponding imaginary lines of the pencils. By Art. [37], displace the pencils so that they are in perspective. Then e and e' will intersect on s, the axis of perspective. The point ee' is then an imaginary point since the vertices of the pencils are the real points on e and e'. But s is a real straight line. Hence it passes through the conjugate imaginary point of ee'. Hence the connectors of this point to the vertices of the pencils, i.e., the conjugate imaginary lines of e and e', are corresponding rays of the pencils.

Correlatively it may be proved that *if two imaginary points are corresponding points of two projective ranges with three pairs of corresponding real points, their conjugate imaginary points are also corresponding points of the projective ranges.*

14. The triangle.

The different types of triangles which may occur are as follows, viz.:

(1) *Real Triangle.* A triangle consisting of three real vertices determining three real sides.

Real Triangle. A triangle consisting of three real lines determining three real vertices.

(2) *Semi-real Triangle.* A triangle consisting of one real vertex and two conjugate imaginary vertices (on a real line) determining a pair of conjugate imaginary lines (for two sides) and a real line for the third side (on which the pair of conjugate imaginary vertices are situated).

Semi-real Triangle. A triangle consisting of one real side and two conjugate imaginary sides (meeting in a real point) determining a pair of conjugate imaginary vertices (lying on the real side) and one real vertex (being the point of intersection of the pair of conjugate imaginary lines which form a pair of sides).

(3) *An imaginary Triangle,* type (*a*). A triangle consisting of one real vertex and two imaginary points situated on a real line determining one real side (on which the imaginary vertices are situated) and a pair of imaginary sides passing through the real vertex.

An imaginary Triangle, type (*a*). A triangle consisting of one real side and a pair of imaginary sides passing through a real point, determining one real vertex (through which the pair of imaginary sides pass) and a pair of imaginary vertices on the real side.

(4) *An imaginary Triangle,* type (*b*). A triangle consisting of a pair of conjugate imaginary vertices and a third imaginary vertex, determining one real side (joining the pair of conjugate imaginary vertices) and a pair of imaginary sides.

An imaginary Triangle, type (*b*). A triangle consisting of a pair of conjugate imaginary sides and a third imaginary side, determining one real vertex (being the point of intersection of the conjugate imaginary sides) and a pair of imaginary vertices.

(5) *An imaginary Triangle,* type (*c*). A triangle consisting of three imaginary vertices, determining three imaginary sides.

An imaginary Triangle, type (*c*). A triangle consisting of three imaginary sides, determining three imaginary vertices.

In all the above cases, except case (4), the triangle and that given by the correlative construction are of the same nature.

15. Extension of Menelaus' theorem.

I. *If an imaginary straight line meets one side of a real triangle in a real point and the other two sides in imaginary points, the product of the ratios of these points, with respect to the triangle, is unity.*

This may at once be deduced from Art 10. For if a real straight line be drawn through S to meet s and s_1 in A and A_1, the triangle OAA_1 will be a real triangle and the imaginary straight line SEE_1 is a straight line which meets two of the sides in imaginary points E and E_1, and the third side in a real point S.

It follows from Art. 10 that

$$(OAEW') = (OA_1E_1\infty).$$

Therefore
$$\frac{OE}{AE} : \frac{OW'}{AW'} = \frac{OE_1}{A_1E_1}.$$

But $\dfrac{OW'}{AW'} = \dfrac{A_1S}{AS}$, by similar triangles.

Therefore
$$\frac{OE}{AE} \cdot \frac{AS}{A_1S} \cdot \frac{A_1E_1}{OE_1} = 1.$$

The theorem may also be proved independently as follows:

Take a real triangle ABC. Let A_b and A_c be the conjugate points of B and C in an involution (with imaginary double points) on BC.

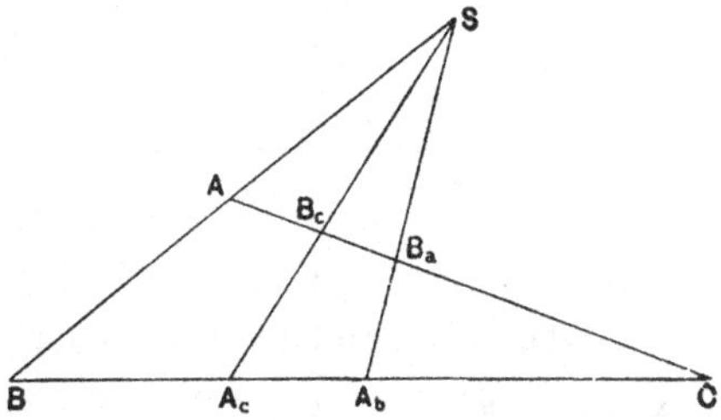

Take any real point S on AB. From S project A_c and A_b into B_c and B_a.

Then the involution B, A_b, C, A_c is projected from S into A, B_a, C, B_c.

Let the ratios of A_c, A_b be a' and a and of B_c, B_a be b' and b. Then the ratios, x and y, of the double points of the involutions are given by

$$x^2 - 2xa' + aa' = 0$$
and
$$y^2 - 2yb + bb' = 0. \qquad \text{(See Art. [61] Ex. (7).)}$$

If c be the ratio of S then

$$cab = 1 \text{ and } ca'b' = 1 \dots\dots\dots\dots\dots\dots\dots(1)$$

The double points are given by

$$(x - a')^2 = a'(a' - a)$$
$$(y - b)^2 = b(b - b'),$$
$$\therefore x = a' \pm \sqrt{a'(a' - a)}, \ y = b \pm \sqrt{b(b - b')}.$$

But from (1) $\dfrac{a}{a'}=\dfrac{b'}{b}=\dfrac{1}{ca'b}=\lambda$ suppose,

$$\therefore \ x=a'\pm\sqrt{a'^2(1-\lambda)},\quad y=b\pm\sqrt{b^2(1-\lambda)}.$$

Let $x_1=a'(1+\sqrt{1-\lambda}),\ y_1=b(1-\sqrt{1-\lambda});$

$$\therefore \ x_1y_1=a'b(1-1+\lambda)=a'b\lambda=\dfrac{a'b}{ca'b},$$

$$\therefore \ x_1y_1c=1.$$

Similarly if $x_2=a'(1-\sqrt{1-\lambda}),\ y_2=b(1+\sqrt{1-\lambda}),$

then $x_2y_2c=1.$

II. *If an imaginary straight line meets the sides of a real triangle in three imaginary points, then the product of the ratios of these points with respect to the triangle is unity.*

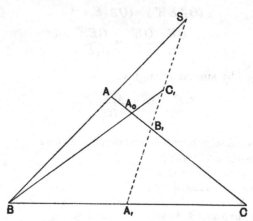

Take any real triangle ABC. Draw through B any real transversal to meet AC in A_0 and the imaginary line SB_1A_1 in C_1, where S is real.

Let the ratios of A_1, B_1, C_1 referred to the triangle A_0BC be x, y, z.

Let $\dfrac{AS}{BS}=c.$

Then from the triangle ABC

$$c\,.\,x\,.\,\frac{CB_1}{AB_1}=1 \ \text{ or } \ c\,.\,x\,.\,\frac{CB_1}{A_0B_1}\,.\,\frac{A_0B_1}{AB_1}=1 \ \text{ or } \ c\,.\,x\,.\,y\,.\,\frac{A_0B_1}{AB_1}=1,\ldots\text{(i)}$$

and from ABA_0

$$c\,.\,\frac{BC_1}{A_0C_1}\,.\,\frac{A_0B_1}{AB_1}=1 \ \text{ or } \ c\,.\,\frac{1}{z}\,.\,\frac{A_0B_1}{AB_1}=1. \qquad\ldots\ldots\ldots\text{(ii)}$$

Dividing the expressions (i) and (ii)

$$\frac{1}{xy} = z,$$

$$\therefore \quad xyz = 1.$$

Hence the imaginary line $A_1B_1C_1$ meets the sides of the triangle of reference A_0BC in three points (imaginary) for which Menelaus' theorem holds.

16. Extension of Ceva's theorem.

I. *If the connector of any imaginary point to one of the vertices of a real triangle is real, the product of the ratios of the three points in which its connectors to the vertices meet the opposite sides of the triangle is* -1.

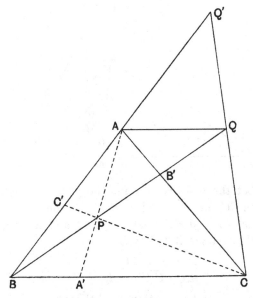

Let P be the imaginary point and let AP, BP, CP, meet the opposite sides of the triangle in A', B', C'. Let BPB' be real. Draw AQ parallel to BC to meet BB' in Q. Join CQ to meet AB in Q'.

Then $\quad\quad\quad (A \cdot BB'PQ) = (BCA'\infty).$

But $\quad\quad\quad (C \cdot BB'PQ) = (BAC'Q').$

Therefore $\quad\quad (BCA'\infty) = (BAC'Q').$

$$\therefore \quad \frac{BA'}{CA'} = \frac{BC''}{AC''} : \frac{BQ'}{AQ'}.$$

Therefore $\dfrac{BC'}{AC'}\dfrac{CA'}{BA'}=\dfrac{BQ'}{AQ'}=\dfrac{BC}{AQ}=\dfrac{CB'}{B'A}.$

Therefore $\dfrac{BC'}{AC'}\dfrac{CA'}{BA'}\dfrac{AB'}{CB'}=-1.$

II. *The connectors of an imaginary point to the vertices of a real triangle meet the opposite sides in three points the product of whose ratios with reference to the triangle is* -1.

Let P be the imaginary point and let the connectors of this point with the vertices of the triangle ABC meet the sides in A B', C',

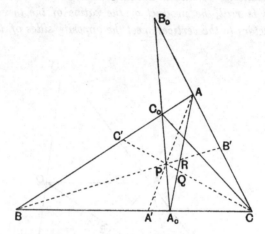

respectively. Let the real line through the point P meet the sides in A_0, B_0, C_0. Join CC_0 and AA_0. Let a, b, c be the ratios of A', B', C' respectively. Let CP and BP meet AA_0 in Q and R respectively.

Then from the triangle AA_0C by I.

$$\frac{AQ}{A_0Q}\cdot\frac{A_0A'}{CA'}\ \frac{CB_0}{AB_0}=-1,$$

and from the triangle AA_0B by Menelaus' theorem

$$\frac{A_0Q}{AQ}\frac{BC}{A_0C}\frac{AC'}{BC'}=+1.$$

Therefore $\qquad c\cdot\dfrac{A_0A'}{CA'}\dfrac{CB_0}{AB_0}\cdot\dfrac{BC}{A_0C}=-1.$ ....(I)

Similarly from the triangle AA_0B by I.

$$\frac{A_0R}{AR}\cdot\frac{AC_0}{BC_0}\frac{BA'}{A_0A'}=-1,$$

and from the triangle AA_0C by Menelaus' theorem

$$\frac{AR}{A_0R} \cdot \frac{A_0B}{CB} \cdot \frac{CB'}{AB'} = +1.$$

Therefore $\quad b \cdot \frac{BA'}{A_0A'} \cdot \frac{AC_0}{BC_0} \cdot \frac{A_0B}{CB} = -1$ ....(II)

Hence multiplying together (I) and (II)

$$-b \cdot c \cdot \frac{BA'}{CA'} \cdot \frac{CB_0}{AB_0} \cdot \frac{AC_0}{BC_0} \cdot \frac{A_0B}{A_0C} = +1.$$

Hence $a \cdot b \cdot c = -1$.

Therefore Ceva's theorem holds for a real triangle.

It may be shown as follows from the first case of Menelaus' theorem, which has been proved independently of this result (Art. 15), that the anharmonic ratios of two partly imaginary ranges obtained by projection from a real point are equal.

Let two imaginary lines be drawn through a real point C' on the side AB of a real triangle ABC to meet the two other sides in B', A' and B'', A''

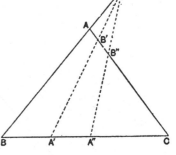

Let the ratios of C', B', B'' be

$$c, \ b'+ib_1', \ b''+ib_1'',$$

respectively.

Then the ratios of A' and A'' (by the extension of Menelaus' theorem) are

$$\frac{1}{c(b'+ib_1')} \text{ and } \frac{1}{c(b''+ib_1'')}.$$

Therefore $\quad (BCA'A'') = \dfrac{BA'}{CA'} : \dfrac{BA''}{CA''}$

$$= \frac{1}{c(b'+ib_1')} : \frac{1}{c(b''+ib_1'')}$$

$$= (b''+ib_1'') : (b'+ib_1')$$

$$= \frac{CB''}{AB''} : \frac{CB'}{AB'}$$

$$= (ACB'B'').$$

It does not follow, and it is not true, that the real portions of the determining elements of points on one range are projected into the real portions of the determining elements of points on the other range. This is only true when the cutting lines are parallel.

17. The quadrangle and quadrilateral.

There are various kinds of quadrangles and quadrilaterals which differ with the nature of the four points and the four straight lines which determine them.

The most important of these are the real quadrangle and quadrilateral and the semi-real quadrangle and quadrilateral. The latter are constructed as follows:

Semi-real Quadrangle.

1*st Kind.*

Two pairs of conjugate imaginary points (A, A' and B, B') determine

(1) a pair of real lines and two pairs of conjugate imaginary lines;

(2) three real points, being the points of intersection of the three pairs of lines.

Semi-real Quadrilateral.

Two pairs of conjugate imaginary lines (a, a' and b, b') determine

(1) a pair of real points and two pairs of conjugate imaginary points;

(2) three real lines, being the connectors of the three pairs of points.

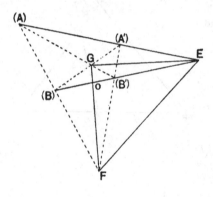
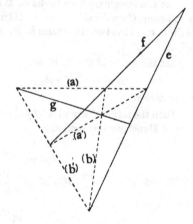

In the figure the lines
   AA', BB' are real,
   AB', BA' are conjugate imaginary lines,
   AB, $A'B'$ are conjugate imaginary lines.
   E, F, G are real, since the intersections of conjugate imaginary lines are real, and are the diagonal points of the quadrangle.

In the figure the points
   aa', bb' are real,
   ab', ba' are conjugate imaginary points,
   ab, $a'b'$ are conjugate imaginary points.
   e, f, g are real, since the connectors of conjugate imaginary points are real, and are the diagonals of the quadrilateral.

2nd Kind.

A pair of imaginary conjugate points (A, A') and a pair of real points (B, C) determine

(1) a pair of real lines and two pairs of imaginary lines;

(2) three points, one of which is real, the other two conjugate imaginary points.

A pair of conjugate imaginary lines (a, a') and a pair of real lines (b, c) determine

(1) a pair of real points and two pairs of imaginary points;

(2) three lines, one of which is real, and the other two a pair of conjugate imaginary lines.

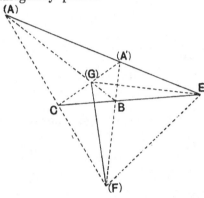
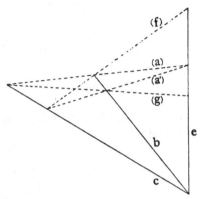

In the figure
AA' and CB are real,
AB and $A'C$ are imaginary,
$A'B$ and AC are imaginary.
The point E is real.

The lines BA, BA' and CA, CA are pairs of conjugate imaginary lines and therefore their points of intersection G and F are conjugate imaginary points. Hence the line GF is real and EG, EF are conjugate imaginary lines.

If in the previous case the pair of conjugate imaginary points B and B' and the pair of real points G and F are looked upon as the determining points, the diagonal points triangle is $AA'O$.

In the figure
aa' and bc are real,
ab and $a'c$ are imaginary,
$a'b$ and ac are imaginary.
The line e is real.

The points ba, ba' and ca, ca' are pairs of conjugate imaginary points and therefore their connectors g and f are conjugate imaginary lines. Hence the point gf is real and eg, ef are conjugate imaginary points.

If in the previous case the pair of conjugate imaginary lines b and b' and the pair of real lines g and f are looked upon as the determining lines, the diagonal triangle consists of a, a' and the connector of gf to bb'.

For the construction of the diagonal points triangle of a semi-real quadrangle and of the diagonal triangle of a semi-real quadrilateral, see Art. [60].

18. Harmonic property of a semi-real quadrangle.

In a semi-real quadrangle the ranges determined on a real side or on a real side of the diagonal points triangle are harmonic.

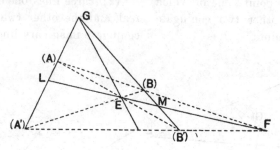

Case I. Let A, A' and B, B' be two pairs of conjugate imaginary points and let

$$AA' . BB' \text{ be } G,$$
$$AB' . BA' \text{ be } E,$$
$$AB . A'B' \text{ be } F.$$

Let EF meet AA' and BB' in L and M.

The ranges G, L, A, A' and G, M, B', B are in perspective with the real point E as centre of perspective.

Therefore $(GLAA') = (GMB'B).$

Similarly the ranges G, M, B', B and G, L, A', A are in perspective with the real point F as centre of perspective.

Therefore $(GMB'B) = (GLA'A).$

Hence $(GLAA') = (GLA'A)$ and therefore the range $GLAA'$ is harmonic.

Since E and F are the centres of perspective of the involutions on GL and GM the range $LMEF$ is harmonic.   (Art. [60].)

Case II. The real points E, F and the pair of conjugate imaginary points B, B' determine a semi-real quadrangle of the second kind in which the ranges on the real lines are as above harmonic.

The harmonic property of a semi-real quadrilateral may be proved in a similar manner.

19. Involution property of a semi-real quadrangle.

The three pairs of opposite sides of a semi-real quadrangle are cut by any real transversal in three pairs of conjugate points of an involution.

Case I. Let the quadrangle be determined by a pair of real points A and B and a pair of conjugate imaginary points A' and B'. Then in the figure the points G and F are a pair of conjugate imaginary points.

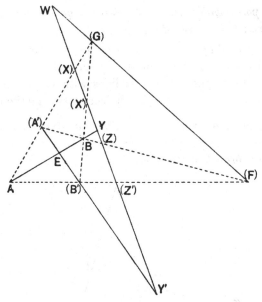

Let the transversal meet the pairs of opposite sides of the quadrangle as in the figure in XX', YY', ZZ'. The point E is a real point. Projecting the range $A'B'EY'$ from the real points A and B it follows that

$$(XZ'YY') = (ZX'YY') = (X'ZY'Y).$$

Hence (Art. 7), XX', YY', ZZ' form an involution.

By Art. 8 the involution is real.

Case II. The quadrangle determined by $A'B'GF$ is a quadrangle determined by two pairs of conjugate imaginary points. Let GF meet the transversal in W.

By Case I the quadrangle A, B, G, F, determines an involution XZ, $X'Z'$, YW on the transversal, therefore $(XZX'Y) = (ZXZ'W)$.

By Case I the quadrangle A, B, A', B' determines an involution XX', YY', ZZ' on the transversal, therefore $(XZX'Y) = (X'Z'XY')$.

Hence $(X'Z'XY') = (ZXZ'W)$.

Therefore ZX', XZ', $Y'W$ form an involution. But these are the three pairs of points in which the opposite sides of the quadrangle $A'B'GF$ are met by the transversal. Hence the theorem is proved.

By Art. 8 this involution is real and has real double points.

The involution property of a semi-real quadrilateral may be proved in a similar manner.

Conjugate points with respect to a semi-real quadrangle and conjugate lines with respect to a semi-real quadrilateral.

From the preceding the construction for the conjugate points of real points with respect to a semi-real quadrangle may be deduced as in Art. [57]. The conjugates of real lines with respect to a semi-real quadrilateral may be constructed by the correlative method.

20. *Any two real involutions of different kinds in the same plane are in imaginary plane perspective, the real branch of each corresponding to the imaginary branch of the other.*

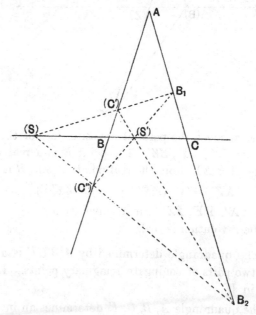

Let B_1 and B_2 be the real double points of one involution and C' and C''' the conjugate imaginary points which are the double points of the other involution.

Then, from the properties of the semi-real quadrangle $B_1B_2C'C'''$, the points S and S' are a pair of conjugate imaginary points and the line SS' is real.

Take the real triangle ABC as triangle of reference.

Let P', P'' be two real points on AC, whose ratios p' and p'' are such that $p'p'' = K^2$, K^2 being the constant of the involution of AC. Let $-K'^2$ be the constant of the involution on AB. Then the ratios of S and S' are

$$+ \frac{1}{\sqrt{-K^2K'^2}} \quad \text{and} \quad - \frac{1}{\sqrt{-K^2K'^2}}. \quad \text{(Art. [60].)}$$

Let the imaginary lines SP' and SP'' meet AB at Q' and Q''

Then the ratios of Q' and Q'' are

$$\frac{\sqrt{-K^2K'^2}}{p'} \quad \text{and} \quad \frac{\sqrt{-K^2K'^2}}{p''} \text{ (by Menelaus' theorem)}.$$

Therefore the product of the ratios of Q' and Q'' is

$$\frac{-K^2K'^2}{p'p''} = -K'^2.$$

Therefore Q', Q'' are a pair of imaginary conjugate points of the involution on AB.

21. Projective ranges and pencils.

(a) *If a range of points, real and imaginary, situated on a real straight line be projected from an imaginary point upon a real straight line the two ranges so obtained are equianharmonic.*

If a pencil of rays, real and imaginary, passing through a real vertex, be cut by an imaginary transversal and the range so found be projected from a real point, the two pencils so formed are equianharmonic.

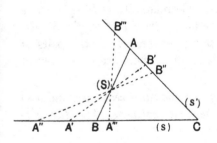

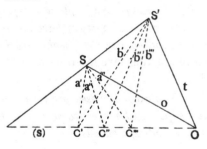

Let the range of points situated on a real line s be $C, A', A'', A'''....$ Let S be the imaginary point which is the centre of projection. Draw BA the real line through S to meet s in B. Let the projections of A', A'', A''', $B...$ on s', be B', B'', B''', $A,...$ and let the ratios of the points $A', A'', A'''...$ with respect to BC be $a', a'', a'''...,$ those of $B', B'', B'''...,$ with respect to CA, $b', b'', b'''...,$ and that of S with respect to AB, c.

Then by Art. 15

$$a'b'c = 1 = a''b''c,$$

$$\therefore \frac{a'}{a''} = \frac{b''}{b'},$$

$$\therefore (BCA'A'') = (CAB''B')$$
$$= (ACB'B''),$$

so $\quad (BCA'A''') = (ACB'B'''),$

$\therefore (CA'A''A''') = (CB'B''B'''),$

$\therefore (A'A''A'''A'''') = (B'B''B'''B'''').$

Let the pencil of rays through the real point S be $o, a', a'', a'''....$ Let s be the imaginary line which cuts them in $O, C', C'', C'''...,$ O being the real point on this line. Let the connectors of $O, C', C'', C'''...$ to S' be $t, b', b'', b'''....$ Take $OS'S$ as triangle of reference. Let the ratios of the intersections of $a', a'', a'''...$ with OS' be $a_1', a_1'', a_1'''...$ and the ratios of the intersections of $b', b'', b'''...$ with OS be $b_1', b_1'', b_1'''...$ and the ratio of the point of intersection of OC' with SS' be s_1.

Then $a_1'b_1's_1 = -1 = a_1''b_1''s_1$ by Art. 16,

$$\therefore \frac{a_1'}{a_1''} = \frac{b_1''}{b_1'},$$

therefore as on the left-hand side the ranges determined by the pencils on the real transversals OS' and OS are equianharmonic and therefore the pencils are equianharmonic.

If a system of real and imaginary points on a real straight line be joined to a real point (Art. 10) or to an imaginary point (Art. 21) the pencil so formed is cut by real transversals in projective ranges. *Two pencils so cut by real transversals in projective ranges are said to be projective.*

As in Art. [34] it can be shown that *if two projective ranges on real bases have the point of intersection of their bases for a self-corresponding point, the ranges are in plane perspective.* Correlatively, *if two projective pencils with real vertices have the connector of their vertices for a self-corresponding ray the pencils are in plane perspective,* i.e. the pairs of corresponding rays intersect on a straight line. These are extensions of Art. 13.

(b) *Two superposed projective ranges on the same real base, and also two superposed projective pencils with the same real vertex, have two self-corresponding elements.*

Consider two superposed ranges determined by points A, B, C and A', B', C', real or imaginary. Let V be the point of the first, which corresponds to the point at infinity of the second, and W' the point of the second which corresponds to the point at infinity of the first.

Then $$(ABV\infty) = (A'B'\infty\, W'),$$

$$\therefore \frac{AV}{BV} = \frac{B'W'}{A'W'},$$

$$\therefore AV . A'W' = BV . B'W' = \text{constant} = K + iK' \text{ (suppose)}.$$

If P be a self-corresponding point of the ranges, then
$$PV . PW' = K + iK'.$$

Let the distances of V, W' and P from a given point of the base be $a + ia'$, $b + ib'$ and x respectively. Then
$$(x - a - ia')(x - b - ib') = K + iK'.$$

As this is a quadratic equation in x, there are two values of x and therefore two self-corresponding points. Since two superposed projective pencils are cut by a real transversal in two superposed projective ranges, two superposed pencils have two self-corresponding rays.

22. Real involution pencil.

If the pairs of conjugate points, real and imaginary, of a real involution range on a base s be joined to a real or an imaginary point S, the pencil so formed is cut by real transversals in ranges, which are projective with the ranges formed by the corresponding points of the given involution (Art. 21). Hence the pencil is cut by all real transversals in involutions and is therefore said to form *an involution pencil*.

If a pair of conjugate points A, A' of the involution on s coincide, their projections on any real transversal coincide, and therefore the double rays of the involution pencil are obtained by joining to the vertex of the pencil the double points of the involution range and are therefore real or imaginary.

If the vertex S of the pencil is real, the involution pencil always has a real branch and is said to form a real involution pencil.

If two pairs of the real conjugate rays of an involution pencil with a real vertex S are at right angles every pair of real conjugate rays of the pencil are at right angles (Art. [58]). If through another real point S' two other pairs of rays, parallel to the rays which meet at S, be drawn they will be at right angles. The rays of the pencil vertex S' being parallel to the rays of the pencil vertex S meet the line at infinity in the same two pairs of points and will therefore determine the same involution on the line at infinity. The double rays of both pencils are obtained by joining the double points of this involution—termed *the circular points at infinity*—to S and S' The lines joining S and S' to the circular points at infinity are termed *the critical lines* of S and S' Real lines, which are at right angles, are therefore harmonic conjugates of the critical lines through their point of intersection. There are also through S and S' pairs of imaginary conjugate rays of the same involution pencils. These pairs of rays are also harmonic conjugates of the critical lines and are defined as *imaginary lines which are at right angles.* Hence, if an imaginary line passes through S, there is through S one imaginary line perpendicular to it, viz. its harmonic conjugate with regard to the critical lines through S.

Similarly, if S be an imaginary point, there are a pair of critical lines through it, which are the connectors of S to the circular points at infinity. Pairs of lines through S which are at right angles are defined as being harmonic conjugates of these critical lines.

If $y=(m+im')x$ be an imaginary line through the origin and $y=Mx$ be the line perpendicular to it, then since these lines are harmonic conjugates of the critical lines, whose equation is $x^2+y^2=0$ (see Art. 78),

$$1+M(m+im')=0,$$

$$\therefore M=-\frac{1}{m+im'}.$$

23. Anharmonic ratio of pencils subtended by points in the same plane at a real or purely imaginary point.

(1) *If four real points A, B, C, D are joined to a real point S the anharmonic ratio of the pencil so formed is real.* This is obvious.

(2) *If two pairs of conjugate imaginary points A, A', B, B' are joined to a real point S, the anharmonic ratio of the pencil so formed is real.*

Let the lines AA', BB' meet at O. Then OA and OA' are of the form $a+ia'$ and $a-ia'$.

From S, B and B' project into a pair of conjugate imaginary points B_1 and B_1' on OAA'. Hence the form of OB_1 and OB_1' is $b+ib'$, $b-ib'$.

Hence the anharmonic ratio of the pencil $(S \cdot AA'BB')$ is

$$\frac{a+ia'-b-ib'}{a-ia'-b-ib'} : \frac{a+ia'-b+ib'}{a-ia'-b+ib'}$$

$$=\frac{(a-b)^2+(a'-b')^2}{(a-b)^2+(a'+b')^2}.$$

Anharmonic Ratios

35

(3) *If a pair of conjugate imaginary points A, A' and a pair of real points B, C be joined to a real point S, the anharmonic ratio $K+iK'$ of the pencil $(S.AA'BC)$ is such that $K^2+K'^2=1$.*

If in the preceding B and B' are a pair of real points they project from S into a pair of real points B_1 and C_1 on OA.

Let
$$(AA'B_1C_1)=K+iK'.$$

Then, since A and A' are conjugate imaginary points,
$$(A'AB_1C_1)=K-iK'.$$

Therefore
$$(AA'B_1C_1)=\frac{1}{(A'AB_1C_1)}=\frac{1}{K-iK'}=K+iK';$$

$$\therefore\ 1=(K-iK')(K+iK')=K^2+K'^2$$

(4) *If four purely imaginary points A, B, C, D are joined to a purely imaginary point S, the anharmonic ratio of the pencil so formed is real.*

This is obvious since the ratio of two purely imaginary quantities is real.

(5) *If two pairs of conjugate imaginary points A, A' and B, B' are joined to a purely imaginary point S, the anharmonic ratio of the pencil so formed is real.*

This result follows from (2) by substituting ia, ia', ib, ib' etc. for a, a', b, b' etc.

When looked at from the purely imaginary point of view the determining quantities for a pair of conjugate imaginary points are $ia-b$ and $ia+b$, i.e. the imaginary parts are the same and they differ in the sign of the real part.

(6) *If a pair of conjugate imaginary points A, A' and a pair of purely imaginary points B, C be joined to a purely imaginary point S, the anharmonic ratio $K+iK'$ of the pencil $(S.AA'BC)$ is such that $K^2+K'^2=1$.*

The values of $(AA'B_1C_1)$ and $(A'AB_1C_1)$ will in this case be the same as in case (3) except that all the quantities determining the position of the points are multiplied by i. This quantity will divide out in the expressions for the anharmonic ratios and therefore K and K' are as in (3), i.e. $K^2+K'^2=1$.

A quantity $K+iK'$, where $K^2+K'^2=1$, can be expressed as the quotient of two conjugate imaginary quantities.

Let
$$K+iK'=\frac{A+iB}{A-iB}=\frac{(A^2-B^2)+2iAB}{A^2+B^2}.$$

Then if
$$K=\frac{A^2-B^2}{A^2+B^2}\text{ and }K'=\frac{2AB}{A^2+B^2},\quad K^2+K'^2=1.$$

But
$$K(A^2+B^2)=(A^2-B^2),\quad\therefore\ \frac{A^2}{B^2}=\frac{1+K}{1-K}.$$

Since $K<1$, because $K^2+K'^2=1$, this expression is positive whether K is positive or negative. Therefore $\dfrac{A}{B}=\sqrt{\dfrac{1+K}{1-K}}$ is real.

Hence
$$K+iK'=\frac{\sqrt{1+K}+i\sqrt{1-K}}{\sqrt{1+K}-i\sqrt{1-K}}.$$

24. *Condition that the anharmonic ratio of four collinear points may be real.*

Let the four points A, B, C, D be determined by distances $a+ia'$, $b+ib'$, $c+ic'$, $d+id'$ measured from an origin on the straight line.

Let
$$A \equiv \frac{a-c}{a'-c'}, \quad B \equiv \frac{b-c}{b'-c'}, \quad C \equiv \frac{a-d}{a'-d'}, \quad D \equiv \frac{b-d}{b'-d'}.$$

Then $(ABCD) = \left\{\frac{a'-c'}{b'-c'} : \frac{a'-d'}{b'-d'}\right\} \cdot \left\{\frac{A+i}{B+i} : \frac{C+i}{D+i}\right\} = K+iK'$ (suppose).

If $K' \equiv 0$ and M' represents the expression in the first bracket
$$M'(A+i)(D+i) = K(B+i)(C+i).$$

Hence $M'(AD-1) = K(BC-1)$ and $M'(A+D) = K(B+C)$.

Therefore $A \cdot B \cdot C \cdot D \cdot \left\{\frac{1}{C} + \frac{1}{B} - \frac{1}{D} - \frac{1}{A}\right\} = B+C-A-D$ ....(1)

Hence it may be shown that the required condition is
$$\Delta = \begin{vmatrix} 1 & 1 & 1 & 1 \\ a & b & c & d \\ a' & b' & c' & d' \\ a^2+a'^2 & b^2+b'^2 & c^2+c'^2 & d^2+d'^2 \end{vmatrix} = 0.$$

25. *Relations connecting the anharmonic ratios of collinear imaginary points.*

In the general case with the notation of Art. 24
$$M'(A+i)(D+i) = (K+iK')(B+i)(C+i).$$

Therefore $M'(AD-1) = K(BC-1) - K'(B+C)$

and $M'(A+D) = K(B+C) + K'(BC-1)$.

Hence it may be proved that

$$K' = \frac{\Delta}{\{(b-c)^2+(b'-c')^2\}\{(a-d)^2+(a'-d')^2\}} = \frac{\Delta}{f(b,c) \cdot f(a,d)} \text{ (suppose)} ...(2)$$

and $$K^2+K'^2 = \frac{f(b,d) \cdot f(a,c)}{f(b,c) \cdot f(a,d)}. \quad ....(3)$$

Hence $$K = \pm \frac{1}{f(b,c) \, f(a,d)} \sqrt{f(b,c) f(a,d) f(b,d) f(a,c) - \Delta^2}. \quad ....(4)$$

If $K' = 0$ the anharmonic ratio of the four points and their conjugate imaginary points are equal and
$$K^2 = \frac{f(a,c) \cdot f(b,d)}{f(b,c) \cdot f(a,d)}.$$

This is the result obtained by making Δ zero in (4).

Let $f(a,b) \cdot f(c,d) \equiv S$, $f(a,c) \cdot f(b,d) \equiv S_1$, $f(a,d) \cdot f(b,c) \equiv S_2$.

Then $$K = \frac{S_1+S_2-S}{2S_2}, \quad K'^2 = \frac{2S_1S_2+2SS_1+2SS_2-S_1^2-S_2^2-S^2}{4S_2^2},$$

and since $$K' = \frac{\Delta}{S_2}, \quad 4\Delta^2 = 2S_1S_2+2SS_1+2SS_2-S_1^2-S_2^2-S^2.$$

Hence it is seen that

$$(ABCD) = \frac{S_1 + S_2 - S}{2S_2} + i\frac{\Delta}{S_2}$$

$$(ABDC) = \frac{S_1 + S_2 - S}{2S_1} - i\frac{\Delta}{S_1}$$

$$(ACBD) = \frac{S + S_2 - S_1}{2S_2} - i\frac{\Delta}{S_2}$$

$$(BACD) = \frac{S_1 + S_2 - S}{2S_1} - i\frac{\Delta}{S_1}$$

$$(BADC) = \frac{S_1 + S_2 - S}{2S_2} + i\frac{\Delta}{S_2}$$

and so on, where $4\Delta^2 = 2SS_1 + 2SS_2 + 2S_1S_2 - S^2 - S_1{}^2 - S_2{}^2$.

If $a' = b' = c' = d' = 0$ the corresponding formulae for real points are obtained.

26. Projection from a real point on a real plane.

(a) *Any pair of real points and any pair of conjugate imaginary points can by a real projection be projected into any pair of real points and any pair of conjugate imaginary points, if the two systems are each coplanar.*

Let A, B, C_1 and C_2 be a pair of real points and a pair of conjugate imaginary points in a plane σ and A', B', C_1' and C_2' any pair of real points and any pair of conjugate imaginary points in a plane σ'.

Take any real plane σ'' through A' and any centre of projection (real) S on AA'. Project the figure in the plane σ from S on σ''. Thus a figure in plane σ'' is obtained, viz. A', B'', C_1'', C_2'' in which C_1'' and C_2'' are a pair of conjugate imaginary points.

Let $A'B'$ and $C_1'C_2'$ meet at E' and let $A'B''$ and $C_1''C_2''$ meet at E''. Then E' and E'' are real points.

Since $A'B'E'$ and $A'B''E''$ are coplanar $B'B''$ and $E'E''$ meet at a real point S'. Project the figure in σ'' from S' upon a plane σ''' through the line $A'B'E'$. Then a figure A', B', E', C_1''', C_2''' is obtained in the plane σ'''.

Since C_1', C_2' and C_1''', C_2''' are two pairs of conjugate points in a plane ($C_1'C_2'$ and $C_1'''C_2'''$ are concurrent at E') therefore $C_1'C_1'''$ and $C_2'C_2'''$ meet at a real point S''.

Projecting the figure in σ''' from S'' on σ' the figure $A'B'E'C_1'C_2$ is obtained. Hence the first system of points has been projected into the second.

It should be noticed that C_1 may be projected into C_1' or C_2', and the point C_2 into the other point.

The preceding is equivalent to proving that any two points and an involution in one plane may be projected into any two given points and a given involution in another plane.

(*b*) *To project two pairs of conjugate imaginary points in one plane into any two pairs of conjugate imaginary points in another plane.*

Let A_1, A_2 and B_1, B_2 be any two pairs of conjugate imaginary points in the plane σ and A_1', A_2' and B_1', B_2' any two pairs of conjugate imaginary points in the plane σ'.

Let $A_1 A_2$ and $B_1 B_2$ meet at E and $A_1' A_2'$ and $B_1' B_2'$ meet at E'. Then E and E' are real. Take a centre of projection S on EE' and project the figure in the plane σ upon a plane σ'' which passes through E'

Then A_1, A_2 become a pair of conjugate imaginary points A_1'', A_2'' in σ'' and B_1, B_2 become a pair of conjugate imaginary points B_1'', B_2'' in σ'' and E becomes E'.

Since B_1'', B_2'' and B_1', B_2 meet at E' they are in a plane and the lines $B_1'B_1''$ and $B_2'B_2''$ being conjugate imaginary lines in this plane meet at a real point S'. Take a plane σ''' through the line $B_1'B_2'E'$ and project the figure in the plane σ'' from S' upon this plane.

Then A_1'', A_2'', collinear with E', become a pair of conjugate imaginary points A_1''', A_2''' in σ''' collinear with E'; B_1'', B_2'' become the pair of conjugate imaginary points B_1', B_2' and E' remains the point E'.

Since A_1''', A_2''' and A_1', A_2' are collinear with E' they lie in a plane and therefore the conjugate imaginary lines $A_1'''A_1'$ and $A_2'''A_2'$ meet in a real point S''.

If the figure in the plane σ''' be projected from S'' upon the plane σ', the pair of conjugate imaginary points A_1''', A_2''' are projected into the points A_1', A_2'.

Hence the required real projection has been performed.

27. Semi-real square.

A semi-real square of the first kind is determined by two pairs of conjugate imaginary points situated on two real straight lines, which are at right angles, the distances, imaginary, of the four points from the point of intersection of these lines being equal.

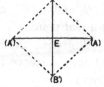

In the figure A and A' and B and B' are the pairs of conjugate imaginary points, and $EA = A'E = EB = B'E$. The lengths of the sides are all equal and are purely imaginary, since $EA^2 + EB^2 = AB^2$. It follows that the triangle BEA is equal in all respects to the triangle AEB' and that the angles BAB', ABA' are right angles. Opposite sides of the figure are conjugate imaginary lines.

A semi-real square of the second kind is determined by a pair of conjugate imaginary points and a pair of real points situated on two real straight lines, which are at right angles, the distances of the pair of imaginary points from the point of intersection of these straight lines being $\sqrt{-1}$ the distances of the real points from the same point.

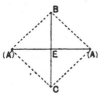

In the figure A and A' are the pair of conjugate imaginary points and B and C the pair of real points,

$$BE = EC = \sqrt{-1}\,.\,A'E = \sqrt{-1}\,.\,EA.$$

The lines CA and CA' and also the lines BA and BA' are pairs of conjugate imaginary lines. The properties of this figure are further discussed in Art. 86.

It follows from the last Art. that *a semi-real quadrangle of the first kind can be projected into a semi-real square of the first kind* and that *a semi-real quadrangle of the second kind can be projected into a semi-real square of the second kind, the projections in both cases being real.*

EXAMPLES

(1) If P and Q be two points on a real straight line and the points R and R' divide the distance between P and Q in the ratios $i\lambda$ and $i\mu$, prove that

$$(PQRR') = \frac{\lambda}{\mu}.$$

(2) Prove that the condition that the four points determined by distances $a + ia'$, $b + ib'$, $a - ia'$, $b - ib'$ should form a harmonic range is

$$(a - b)^2 + (a' + b')^2 + 4a'b' = 0.$$

(3) Prove that no imaginary line can meet a pair of real lines in conjugate imaginary points.

This follows from the fact that the connector of a pair of conjugate imaginary points is a real line.

(4) If the angles between the real rays a, b, c of one pencil are equal to the angles between the corresponding rays a', b', c' of a pencil projective with it, then the angle between any pair of rays p and q of the first is equal to the angle between the corresponding rays p' and q' of the second.

(5) If V and W' are the vanishing points of two real superposed projective ranges of which A and A' are corresponding points so that $VA\,.\,W'A' = K$, prove that the condition that the ranges should have a pair of real self-corresponding points is that $VO^2 + K$ should be positive, where O is the middle point of VW'.

(6) If, in the involution $(ax^2 + 2hx + b) + \lambda\,(a'x^2 + 2h'x + b') = 0$, A, A' and B, B' are the pairs of conjugate points given by $ax^2 + 2hx + b = 0$ and $a'x^2 + 2h'x + b' = 0$ and λ_1 and λ_2 are the values of λ corresponding to the double points of the involu-

tion, then $\lambda_1 + \lambda_2 = 0$ is the condition that $(AA'BB')$ should be harmonic; and generally if $\mu = (AA'BB')$

$$\left(\frac{1+\mu}{1-\mu}\right)^2 = \frac{(\lambda_1 + \lambda_2)^2}{4\lambda_1\lambda_2}$$

From the condition for equal values of x

$$\lambda_1 + \lambda_2 = -\frac{ab' + ba' - 2hh'}{a'b' - h'^2} \text{ and } \lambda_1\lambda_2 = \frac{ab - h^2}{a'b' - h'^2}.$$

$$\therefore \frac{(\lambda_1 + \lambda_2)^2}{\lambda_1\lambda_2} = \frac{\{ab' + ba' - 2hh'\}^2}{(ab - h^2)(a'b' - h'^2)} = 4\left(\frac{1+\mu}{1-\mu}\right)^2 \text{ by Art. [14], Example (3)}$$

CHAPTER II

THE CONIC WITH A REAL BRANCH

28. In the *Principles of Projective Geometry* (Art. [92]) a conic was defined as the locus of the points of intersection of a pair of corresponding rays (real) of two projective pencils (with real vertices). In Art. 95 (*k*) it was proved that such a conic determines on every straight line (real) an involution by means of pairs of collinear conjugate points, and that, when the double points of this involution are real, they are the points of intersection of the line and conic.

The consequences of Axiom I, which are set forth in the preceding chapter, render it possible to enlarge and extend what was proved in the *Principles of Projectine Geometry*. The assumption of Axiom I is that when there are no real double points of a real involution there are a pair of imaginary double points, which are called a pair of conjugate imaginary points, and, as a consequence, a pair of conjugate imaginary lines are defined as the double rays of a real overlapping involution pencil. It follows (Art. 6) that two real projective ranges and two real projective pencils have, in addition to pairs of real corresponding elements, pairs of corresponding imaginary elements, and it is proved that the anharmonic properties of real and of imaginary elements are similar. A pair of self-corresponding elements always exist, when the pencils or ranges are superposed, and (Art. 21 (*b*)) they are either a pair of real, coincident, or conjugate imaginary elements. The same is true of a real involution, which is only a particular case of two superposed projective ranges or pencils.

Hence taking into account the imaginary corresponding elements of two real projective ranges or pencils it is seen that

(1) There are on a conic an infinite number of imaginary points, viz. the points of intersection of pairs of imaginary corresponding elements of the real generating pencils.

(2) On every real straight line in its plane the conic determines a real involution of which the double points are a pair of real, co-incident, or conjugate imaginary points, and these are the points of intersection of the real line and the conic. Consequently every real line in the plane of a real conic meets the conic in a pair of points

real, coincident or conjugate imaginary, and, if a conic passes through an imaginary point it also passes through the conjugate imaginary point.

(3) Every line through a given real point meets its polar in a point, such that this point, the given point and the points of intersection of the line with the conic form a harmonic range.

Similarly the theorems correlative to the above may be shown to be true.

In the extension obtained in this manner it must be borne in mind that the vertices of the pencils and likewise the bases on which the ranges are situated must be real as well as the anharmonic ratios of both ranges and pencils. In Chapter IV the fundamental theorems for the conic are proved for the conic in general, including of course the case of a conic with a real branch, which in this chapter is termed a real conic. The restriction that the vertices of the pencils and the bases of the ranges considered must be real does not apply to the proofs in Chapter IV.

A circle is only a particular case of a conic, so that the preceding applies to a circle with a real branch.

29. Circle with a real branch.

Construction of the involution determined by a real circle on a straight line.

(*a*) If a line p meets a circle in real points E and F, these points are the double points of the involution determined by the circle on the line. P and P', any pair of conjugate points of the involution, are harmonic conjugates of E and F. If O be the foot of the perpendicular from C, the centre of the circle, on p, O is the conjugate of the point at infinity on p and consequently

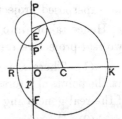

$$OP . OP' = OE^2 = OF^2.$$

A circle described on PP' as diameter cuts the given circle orthogonally (Art. [82]).

If CO meets the circle in R and K, then

$$OR . OK = OE . OF = -OP . OP'$$

(*b*) If a line *p* meets a circle in imaginary points, and the perpendicular from *C*, the centre of the circle, meets *p* in *O*, then *O* is the conjugate in the involution of the point at infinity on *p*, and therefore

$$OP . OP' = \text{constant},$$

where *P* and *P'* are a pair of conjugate points of the involution on *p*.

Let the polar of *P* meet *OC* in *M* and *CP* in *N*; then the triangles *OP'M* and *OCP* are similar,

$$\therefore \frac{OM}{OP'} = \frac{OP}{OC},$$

and

$$OP . OP' = - OM . OC = - OR . OK,$$

since (*OMRK*) is harmonic.

But $OR . OK = OV^2$, where *OV* is the tangent from *O* to the circle. Hence if a circle be described with centre *O* and radius *OV* to meet *OC* in *S*, this circle will cut the given circle at right angles and *PP'*, *QQ'*, ... pairs of conjugate points of the involution will subtend right angles at *S* (Art. [53]).

This circle also determines on *p* the graphs of the double points of the involution, which are the points of intersection of the line and circle.

S is either of the pair of common harmonic conjugates of *MC* and *RK*.

30. Conjugate loci or Poncelet figures for a circle.

By taking the points of intersection of a series of parallel chords with a circle it is possible to obtain a graph of the imaginary portion of a circle. Such a figure may be termed a conjugate locus or a Poncelet figure. Figures of this nature were first given by Poncelet in his *Traité des Propriétés Projectives des Figures*. (See Art. 39.)

Let *O* be the centre of the circle, and consider the points in which lines parallel to any diameter *US* meet the circle.

As long as the distance of these lines from *O* is less than the radius, they meet the curve in the part of the locus which is drawn in a continuous line.

If however a line of the system such as *A'MB'* meets *NON'* in *M* on the sides of *N* and *N'* remote from *O*, the involution determined on this line is overlapping and the double points, in which the circle intersects the line, are a pair of conjugate imaginary points A' and B'. M is the real mean point of A' and B' and the distances MA', MB'

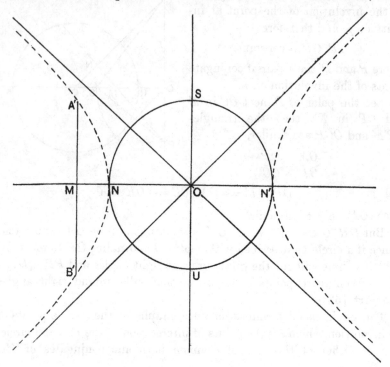

expressed in imaginary units are equal to the tangent from M to the real branch of the circle. Hence, lengths measured parallel to OM being regarded as real and lengths parallel to OS as imaginary, the locus of A' and B' is a rectangular hyperbola, which touches the real branch at N, has O for centre and ON for semi-transverse axis.

For the system of lines parallel to any other diameter there is an exactly equal rectangular hyperbola touching the real part of the curve at the ends of the diameter.

31. Theorems concerning pairs of conjugate imaginary points determined as the points of intersection of a circle and straight line.

(1) *Any pair of conjugate imaginary points may be determined as the points of intersection of a real circle and a real straight line.*

Let M be the centre and P, P' any pair of conjugate points of the involution of which the given pair of conjugate imaginary points are the double points. On

PP' as diameter describe a circle PKP' Draw MC perpendicular to PP' and take
any point C on this line external to the
circle. Join PC to meet the circle described
on PP' as diameter in K. Join KP'. With
centre C and radius, the square of which is
equal to $CK.CP$, describe a circle. This circle
determines the given pair of conjugate imaginary points Q and Q' on PP'. This follows
from the fact that it determines on PP' the
involution of which M is the centre and P, P'
are a pair of conjugate points. The two circles
obviously cut orthogonally since $CU^2 = CK.CP$.

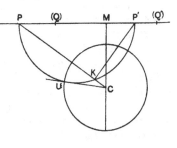

If Q, Q' are the graphs of a given pair of conjugate imaginary points, and a circle
be described on QQ' as diameter, any circle, with centre on MC the perpendicular
through the mean point of Q and Q', which cuts this circle orthogonally, determines
the given pair of conjugate imaginary points on PP'.   This follows also from
Art. 29.

The circle, centre C, since its centre is on the radical axis of circles described on
the lines joining pairs of conjugate points of the involution as diameters and cuts
one of these circles orthogonally, cuts all circles of this coaxal system orthogonally.
Hence all circles with centres on MC, which cut one circle of the first system orthogonally, form a coaxal system.   If the distance of the limiting points of the first
system from M is $\sqrt{-1}K$, the distance of the limiting points of the second system
from M is K.

This result may be put into a slightly different form in which it is an extension
of that given in Art. 82.

In this form it is as follows:

(2) (a) If two circles cut each other orthogonally, each determines inverse points
upon every diameter of the other.

(b) If one circle passes through inverse points with respect to another, they cut
orthogonally.

For if the circle, centre C, which is orthogonal to the circle PKP', does not
meet PP' in real points, it meets this line in a pair of conjugate imaginary points
which are harmonic conjugates of P and P' and, since PP' is a diameter of the
circle PKP', these points are a pair of imaginary inverse points with respect to
this circle.

The converse follows from the fact, that when the circle, centre C, passes through
the imaginary double points of the involution, it is orthogonal to all circles described
on the lines, joining pairs of conjugate points of the involution, as diameters.

The theorem may also be easily proved from a graphical figure.

(3) To construct the common harmonic conjugates of two pairs of collinear points,
either conjugate imaginary or real, determined as the intersections of a straight line
and a pair of circles.

Let the radical axis of the pair of circles meet the base in O.   The circle with
centre O cutting the given circles orthogonally meets the base in the required points.

This construction fails if O is within the circles. In this case the two pairs of points, say A, A' and B, B', are real and the segments overlap. In this case describe on AA' and BB' as diameters circles intersecting in P. On the perpendicular from P on the given line take any point E outside the circles. The circle with centre E cutting the two circles orthogonally will determine the pair of common harmonic conjugates, which are in this case a pair of conjugate imaginary points.

From the nature of this construction it follows that *there can be only one pair of common harmonic conjugates of the two pairs of points.*

Also *the only case when the common harmonic conjugates of two pairs of collinear points, real or conjugate imaginary, are imaginary, is when both pairs are real and the segments overlap.*

(4) *Given a pair of points* (P, P'), *real or conjugate imaginary, as the points of intersection of a straight line and a circle, to determine those harmonic conjugates of the pair, which have a given mean point M.*

Describe a circle with centre M to cut the given circle orthogonally. This circle determines the required points.

This construction fails if M is inside the given circle, in which case P and P' are real and M is between them. Describe a circle on PP' as diameter (Figure of Art. 31 (1)). Draw MC perpendicular to PP' and with centre C, any point on MC external to the circle, describe an orthogonal circle. This meets PP' in the required points Q and Q' which are imaginary.

32. (1) *If through a real point P a real line be drawn to meet a circle in a pair of imaginary points Q and Q' then $PQ \cdot PQ'$ is equal to the square of the tangent from P to the circle.*

If O be the centre of the circle and M the foot of the perpendicular from O on the real line through P, then the points Q and Q' are at imaginary distances from M equal to $\sqrt{-1}\,MT$, where MT is the tangent from M to the circle.

Hence

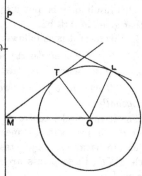

$$PQ \cdot PQ' = (PM - i \cdot MT)(PM + i \cdot MT)$$
$$= PM^2 + MT^2$$
$$= PM^2 + OM^2 - (OM^2 - MT^2)$$
$$= OP^2 - OT^2 = PL^2$$

where PL is the tangent from P to the circle.

Certain important results follow from this theorem.

(a) *If a straight line meet a circle of a coaxal system in a pair of imaginary points, these points are a pair of conjugate points of the involution determined on the line by the system of coaxal circles.*

This follows from the fact that, if the straight line meet the radical axis at P, the squares of the tangents from P to all the circles are equal, and therefore the products of the distances from P of the pairs of points in which the line meets the circles of the system are equal.

(*b*) *The ratio of the tangents from a variable point on a circle to two other circles with which it is coaxal is constant.*

Take P and P' any two points on the circle from which the tangents are drawn. Join PP' to meet the other circles in Q, Q' and R, R'. Then P, P', Q, Q', R, R' are pairs of conjugate points of an involution. Therefore

$$(PP'QR)=(P'PQ'R').$$

Therefore $\dfrac{PQ \cdot PQ'}{PR \cdot PR'} = \dfrac{P'Q \cdot P'Q'}{P'R \cdot P'R'}.$

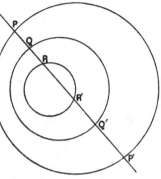

Hence the squares of the tangents from P to the two circles are in the same ratio as the squares of the tangents from P'.

By taking P and P' on the line of centres it follows at once that the ratio of the squares of these tangents is $\dfrac{CC_1}{CC_2}$ where C, C_1 and C_2 are the centres of the circles.

(*c*) *Carnot's theorem holds for the circle even when one or more of the sides of the triangle meets the circle in pairs of imaginary points.*

This can be proved at once as in Art. [89].

(2) *If a chord QPQ' be drawn through a real internal point P to meet a circle in Q and Q', the square of the tangent (imaginary) from P to the circle is equal to $PQ \cdot PQ'$.*

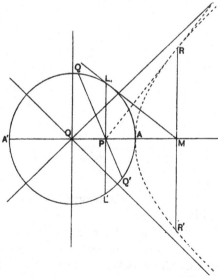

Join O the centre of the circle to P. Through P erect a perpendicular to OP to meet the circle in L and L'. Then $PQ \cdot PQ' = PL \cdot PL' = -PL^2$. Draw RMR' the polar of P, which will meet the circle in a pair of conjugate imaginary points R and R'. Then the square of the tangent PR is PR^2 and

$$PR^2 = PM^2 - RM^2$$
$$= PM^2 - ML^2, \text{ since } LPL' \text{ is the polar of } M$$
$$= -PL^2 = PQ.PQ'.$$

Hence generally *if through a real point P any real chord be drawn to meet a circle in Q and Q', $PQ.PQ'$ equals the square of the tangent from P to the circle.*

To draw a real circle through a real point A and an imaginary point P.

Let P' be the conjugate imaginary point of P. Then the line PP' is real. Let M be the mean point of P and P'. Draw MON perpendicular to PP'. Draw AN, the perpendicular from A on MON.

Construct P_1 and P_1' the graphs of P and P'. On P_1P_1' as diameter describe a real circle. Let CO, the radical axis of this circle and A, meet MN in O. Then O is the centre of the required circle and, if OT be the tangent from O to the circle on P_1P_1', then $OA = OT$ is the radius.

33.   The points of intersection of two circles.

Two circles determine the same involution (a) on their radical axis, and (b) on the line at infinity.

(a) Let P be any point on the radical axis of the circles. Then the tangents from P to the circles are equal and, therefore, a circle

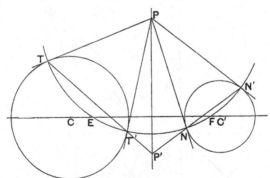

with centre P and radius equal to the length of these tangents cuts the two circles orthogonally.

The chords TT' and NN' are the polars of P with respect to the two circles. These lines are also radical axes of the circles taken in pairs. Hence, since the three radical axes of three circles taken in pairs are concurrent, TT' and NN' meet in a point P' on the radical axis of the first two circles.

Hence the conjugate of P on the radical axis with respect to both the circles is P'. Hence the two circles determine the same involution on their radical axis.

(*b*) Let C and C' be the centres of the circles. Consider a point A at ∞. Its polar with respect to the first circle passes through C, the pole of the line at infinity, and is perpendicular to CA.

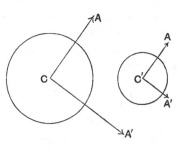

Similarly, the polar of A with respect to the second circle is the line through C' perpendicular to $C'A$. Therefore these polars of A are parallel lines through C and C'. They therefore meet the line at infinity in the same point A'.

Hence the two circles determine the same involution on the line at infinity.

34. (1) *Every circle meets the line at infinity in the same pair of conjugate imaginary points.*

Connected with every pair of circles there are two chords on which the circles determine the same involution. These are their radical axis and the line at infinity. The radical axis is a different line for different pairs of circles, but the line at infinity is the same line. On the line at infinity every circle determines the same involution, namely, the involution obtained by drawing pairs of conjugate (i.e. orthogonal) diameters through the centre. Hence every circle passes through the double points of this involution, which since the involution is an overlapping involution are a pair of conjugate imaginary points. These points are termed the circular points at infinity, or the critical points.

The lines joining the circular points at infinity to the centre of a circle, which is the pole of the line at infinity with regard to the circle, are imaginary tangents to the circle. These tangents are the double rays of the involution pencil made up of pairs of conjugate (or orthogonal) diameters of the circle. They are the critical lines through the centre. (Art. 22.)

(2) *All conics through the circular points at infinity are circles.*

Draw a circle through the circular points at infinity and a conic through them which is supposed not to be a circle. Since these curves intersect the line at infinity in the same pair of points they determine the same involution on it.

Draw parallel diameters of the circle and the conic and their conjugate diameters. These pass through the same pairs of conjugate points of the involution on the line at infinity. In the case of the circle, these pairs of conjugate diameters are at right angles. Therefore the conjugate diameters of the conic are at right angles. Therefore all the pairs of conjugate diameters of the conic are at right angles and consequently it is a circle.

35. *If the self-corresponding rays of two superposed projective pencils are the lines joining the vertex to the circular points at infinity the angle between pairs of corresponding rays is constant.*

Describe a circle through the vertex S to meet the rays of one pencil in A, B $C, \ldots$ and the corresponding rays of the other in A', B', C', ... Join AB' and $A'B$ to meet at K and AC' and $A'C$ to meet at L.

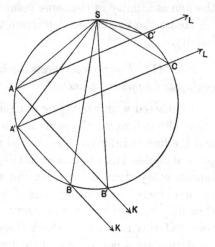

Then the self-corresponding rays of the pencils are the lines joining the points of intersection of KL with the circle to S. (Art. [109].)

If these points are the circular points at infinity, KL must be the line at infinity, and therefore AB' and $A'B$ are parallel, as are also AC' and $A'C$.

Since AB' and $A'B$ are parallel the arcs AA' and BB' are equal. Hence the angles ASA' and BSB' are equal. Similarly the angles between other pairs of corresponding rays are equal to the angle ASA'.

Conversely if the angles ASA', BSB', CSC' are equal, the lines AB', $A'B$; AC', $A'C$; ... will be parallel in pairs and KL will be the line at infinity. This line will meet the circle in the circular points at infinity, and therefore the self-corresponding rays of the pencils are the connectors of the circular points at infinity to S.

36. Every pair of circles intersect in the circular points at infinity and in a pair of points on their radical axis which may be either a pair of real points or a pair of conjugate imaginary points.

Hence their four points of intersection are either:

(1) Two pairs of conjugate imaginary points,

or (2) A pair of conjugate imaginary points and a pair of real points.

These form a semi-real quadrangle of the 1st or 2nd kind (Art. 17).

If in the 1st figure of Art. 17 A, A' are taken as the circular points at infinity, Ω and Ω', and B and B' as the other pair of imaginary points of intersection of the circles, then the line BB' is real, viz. the radical axis.

If in the 2nd figure of Art. 17 A, A' are taken as the circular points at infinity, Ω and Ω', and B and C as the real pair of points of intersection of the circles, then the line BC is real, viz. the radical axis.

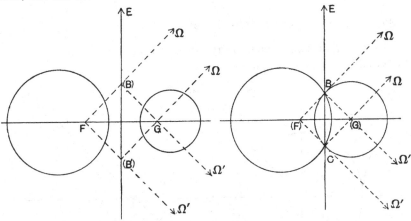

F and G are the real common harmonic conjugates of the two pairs of points in which the line of centres meets the circles. (Art. [84].) They are also harmonic conjugates of the points where the radical axis and the line at infinity meet the line of centres.

E is the real point of intersection (at infinity) of the radical axis and the line at infinity.

EFG is a real triangle.

F and G are a pair of conjugate imaginary points, namely, the common harmonic conjugates of the points where the line of centres meets the circles. (Art. [84].)

E is the real point of intersection (at infinity) of the radical axis and the line at infinity.

The triangle EFG has a real vertex E and a real side FG. F and G are a pair of conjugate imaginary points and the lines EF, EG are a pair of conjugate imaginary lines.

$BCFG$ is a semi-real square of the 2nd kind.

$BB'FG$ is a semi-real square of the 2nd kind.

4—2

Hence *every pair of circles have a common self-conjugate triangle which may be real or semi-real.* The vertices are the common inverse points of the two circles and the point at infinity perpendicular to their common diameter. The circles have also three pairs of common chords, viz. (i) the radical axis and the line at infinity, (ii) the critical lines through one common inverse point, (iii) the critical lines through the other common inverse point.

37. Poncelet figure of two circles.

The fact that two circles always intersect in four points may be illustrated by a Poncelet figure.

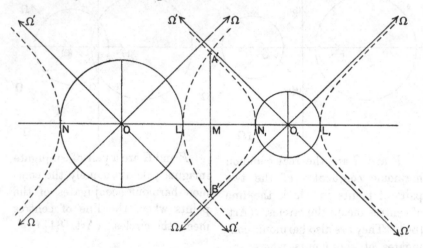

The above Poncelet figure of the curves is constructed by taking as real axis the line joining the centres of the circles. The hyperbolae (rectangular) intersect in two points A' and B' and the lengths MA' and MB' give the imaginary coordinates of the points of intersection, while MO and MO_1 give the real coordinates. $A'B'$ is of course the radical axis of the circles, which is the real connector of the conjugate imaginary points A' and B'.

It is obvious from the figure that the two circles determine the same involution on their radical axis, viz. the involution of which M is the centre and of which the constant is minus the square of the tangent from M to either of the circles.

The principles on which the figure is constructed are explained in Chapter VI.

EXAMPLES

(1) "If two circles cut each other orthogonally each determines inverse points upon every diameter of the other." Prove this in the case where the orthogonal circle meets the diameter in real points.

Give any justification of the extension of this theorem to the case in which the diameter meets the orthogonal circle in imaginary points. (L. U. 1904.)

(2) "We have in the plane a special line, the line at infinity; and on this line two special (imaginary) points, the circular points at infinity. A geometrical theorem has either no relation to the special line and points and it is then *descriptive*; or it has a relation to them and it is then *metrical*."

Explain and comment on this statement. (L. U. 1904.)

(3) If the points P and Q are a pair of conjugate imaginary points, which are also a pair of conjugate points with respect to a circle, prove that the line PQ meets the circle in real points.

(4) Prove that any two straight lines at right angles are harmonic conjugates of the lines joining their point of intersection to the circular points at infinity.

(5) Show that the three poles of a straight line with respect to the three pairs of points of intersection of four given straight lines lie upon another straight line conjugate to the first straight line with respect to each of the three pairs of points.

When the four given lines are the connectors of two given real points with the circular points at infinity, construct the conjugate of a given straight line.

(6) Show that every circle in a given plane may be regarded as passing through the same two imaginary points at infinity.

(7) Prove that no two pairs of conjugate imaginary points can be pairs of harmonic conjugates.

(8) Show that a circle which passes through a real point and a pair of conjugate imaginary points is real.

38. Conic with a real branch.

Construction of the pair of conjugate imaginary points in which a real line meets a conic.

The imaginary points in which a line meets a conic are the double points of the involution which the conic determines on the line.

Let the conic determine on the line l an involution of which KK', LL' are pairs of conjugate points. Let CD be the diameter parallel to l and let its conjugate diameter meet the curve in A and B and l in O'.

Then O the pole of l is on AB.

Hence (Art. [136])

$$O'K \cdot O'K' = O'L \cdot O'L'$$
$$= \mp \frac{CD^2}{CB^2} \cdot O'A \cdot O'B.$$

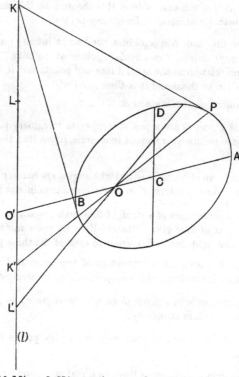

(l)

Therefore if M' and N' are the graphs of the double points of this involution

$$\frac{O'C^2}{CA^2} \mp \frac{O'M'^2}{CD^2} = 1,$$

where N' may be written for M' and the $-$ or $+$ sign must be taken according as the conic is an ellipse or hyperbola. Hence the locus of the graphs of the imaginary double points on systems of parallel chords is a conic touching the given conic at A and B.

The figure so obtained may be termed a Poncelet figure.

39. Conjugate loci or Poncelet figures for a conic.

This result may be illustrated as follows:

(a) *The ellipse.* If l meet the conic in imaginary points M_1' and N_1', and AB, the diameter conjugate to the diameter parallel to l, in O', then

$$-\frac{M_1'O'^2}{CD^2} = \frac{CO'^2}{CB^2} - 1 \dots\dots\dots\dots\dots\dots\dots(\mathrm{i})$$

= a positive quantity, since $CO' > CB$.

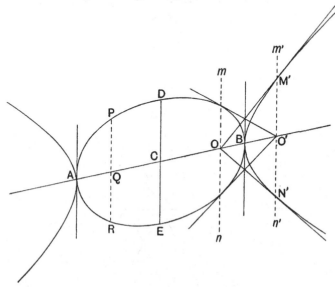

Therefore $M_1'O'$ must be a purely imaginary quantity. Let M' and N' in the figure* be the graphs of the points M_1' and N_1'. Then

$$-\frac{M'O'^2}{CD^2} + \frac{CO'^2}{CB^2} = 1.$$

Hence for a system of chords parallel to l the locus of the graphs is a hyperbola touching the original ellipse at A and B.

(*b*) *The hyperbola.* If in the figure of (*a*) the hyperbola is given it will be found in exactly the same way that the ellipse is the graph of the points of intersection of chords parallel to CD with the hyperbola.

The parabola. If PR is any chord of the parabola, S the focus, and AQ the diameter corresponding to PR, then

$$PQ^2 = 4 \cdot AS \cdot AQ.$$

Draw a line l as in the figure parallel to PR.

---

* This figure is a reproduction of Figure (6) in Poncelet's *Traité des Propriétés Projectives des Figures.* Paris, 1822.

Then if M_1' be on the curve,

$$M_1'O'^2 = 4 \cdot AO' \cdot AS,$$

where AO' is negative.

$$\therefore \ M_1'O'^2 = -4 \cdot O'A \cdot AS.$$

Hence $M_1'O'$ is imaginary. Let M' be its graph. Then

$$M'O'^2 = 4 \cdot O'A \cdot AS.$$

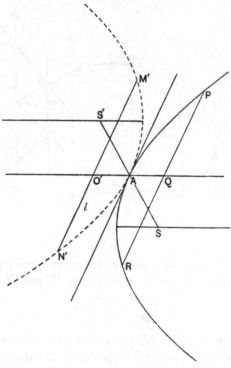

Hence, if a point S' be taken on SA such that $S'A = AS$, the locus of M' for a system of chords parallel to l is a parabola equal to the given one having its focus at S' and touching the given parabola at A. This parabola is the graph of the imaginary points of the parabola for chords parallel to PR.

From the preceding it is seen that a Poncelet figure gives the graphs of the intersections of a conic and a system of real parallel lines, distances measured parallel to one direction being real and those parallel to another direction being purely imaginary. In fig., page 55, the tangent at M' to the graph represents a tangent to the imaginary branch.

It meets AB in a real point O. This is the one real point on the tangent at M'. It is the pole of the line $M_1'N_1'$.

The figure given in this Art. may be obtained from that of Art. 30 by projection from a real point on a parallel plane, since in such a projection real lengths are projected into real lengths and imaginary lengths into imaginary lengths. A more complete figure is given in Art. 127.

40. The following, which is a particular case of the anharmonic property of a conic, may be deduced from Art. [150].

If two fixed real points on a conic and a pair of conjugate imaginary points on the same conic are joined to a variable real point on the conic, the pencil so formed is cut by any real transversal in two real and two imaginary points whose anharmonic ratio is constant.

Let A and B be any two fixed real points on the conic and X_1X_2, Y_1Y_2 any two pairs of conjugate points of the involution determined by the conic on any real line in its plane. If the points A and B are projected from any point S on the conic into the points A_1, B_1 on the real line, then, by Art. [150], for all positions of S

$$\frac{\{(A_1X_1Y_2B_1)-(A_1Y_1X_2B_1)\}^2}{(X_1X_2Y_1Y_2)} \text{ is constant.}$$

Take X_1, X_2 as the double point E, and Y_1, Y_2 as the double point F, of the involution X_1, X_2, Y_1, Y_2. These may be the imaginary points in which the line meets the conic.

Then $\{(A_1EFB_1)-(A_1FEB_1)\}^2$ is constant.

Let $(A_1EFB_1)=\lambda$. Then $\{2\lambda-1\}^2$ is constant.

Therefore λ the anharmonic ratio of the pencil formed by joining the double points of the involution—which may be any pair of conjugate imaginary points on the curve—and the pair of real points—A and B—to any real point on the curve is constant.

In the case of a real conic it is seen that:

(1) *No imaginary line can touch the conic at a real point and no real line can touch the conic at an imaginary point on the conic.*

A tangent at a real point passes through two real points on the curve, and since it passes through two real points it must be a real line.

No real line can touch a conic at an imaginary point, for as it meets the conic in an imaginary point it must meet the conic in the conjugate imaginary point, and an imaginary point and its conjugate can only coincide in a real point.

(2) There is one real point on an imaginary tangent at an imaginary point. If the imaginary point of contact is joined to the conjugate imaginary point on the curve, a real line is obtained. The tangents at this pair of conjugate imaginary points are conjugate imaginary lines and pass through the pole of their connector which being a real point is the one real point on the two imaginary tangents.

(3) If A, A' and B, B' be pairs of conjugate imaginary points in which real chords from G meet a conic, it is seen from the property of the semi-real quadrangle (Art. 17) that $AB' \cdot A'B$ and $AB \cdot A'B'$ are real points on the line joining L and M the harmonic conjugates of G with respect to AA' and BB', i.e. on the polar of G. If the points A and B coincide and likewise A' and B', then the lines AB and $A'B'$ become the imaginary tangents at a pair of conjugate imaginary points and F the pole of GAA' is on the polar of G. This is otherwise obvious.

Consider the inscribed quadrangle A, A', B, B' of which A, A' and B, B' are pairs of conjugate imaginary points and EGF the real diagonal points triangle. The tangents at A and A' intersect in a real point on EF as also do the tangents at B and B'. These real points are the real points on these tangents. The tangents at A and B intersect in an imaginary point, which is the pole of the imaginary line FBA.

Join EG meeting FBA and $FA'B'$ in R and S. Then since
$$(FRAB) = (FSA'B') = -1,$$
RS is the polar of F (Art. 18) and the imaginary tangents at A and B intersect in an imaginary point on the real line EG.

Hence *if through a real point an imaginary line is drawn to meet a conic in a pair of imaginary points, the imaginary tangents at these points intersect in an imaginary point on the real polar of the real point.*

41. Diameters of a conic.

Every real line through the centre of an ellipse or a parabola—the centre of the latter curve being at infinity—meets the curve in real points. This is not however the case with the hyperbola. A diameter may meet the curve in a pair of real points A and A'. It may however meet the curve in a pair of conjugate imaginary points B and B'. These are of course the double points of an overlapping involution the centre of which is the centre of the curve. The lengths CB and CB' are imaginary and $CB^2 = CB'^2 =$ the product of the distances from C of the pair of equidistant conjugate points of the involution, or the product of the distances of any pair of conjugate points of the involution. In Art. [136], where the case of the hyperbola was considered, B and B' were taken as the pair of equidistant conjugate points of the involution on BCB'. If however the imaginary semi-diameter CB be used to determine B, it follows that in that Art. both for the ellipse and the hyperbola

$$XY \cdot XY' = -XA \cdot XA' \cdot \frac{CB^2}{CA^2}$$

and
$$\frac{XP^2}{CB^2} + \frac{CX^2}{CA^2} = 1.$$

42. (i) *If P be a fixed point through which a variable real line PO is drawn and CB be the parallel semi-diameter of an ellipse or hyperbola, and its conjugate semi-diameter CA meets the line OP in O, then $\dfrac{PO^2}{BC^2} + \dfrac{OC^2}{AC^2}$ is constant.*

Describe through P a similar and similarly situated conic and let CA and CB meet this conic in A_1 and B_1. Then $\dfrac{CA}{CA_1} = \dfrac{CB}{CB_1} = \lambda$ (a constant).

Hence
$$\frac{PO^2}{BC^2} + \frac{OC^2}{CA^2} = \frac{1}{\lambda^2}\left\{\frac{PO^2}{B_1C^2} + \frac{OC^2}{A_1C^2}\right\} = \frac{1}{\lambda^2} \text{ by Art. [41].}$$

(ii) *If P be a fixed point through which a variable real line PO is drawn to meet an ellipse or hyperbola in imaginary points Q, Q′, and if CB be the semi-diameter parallel to the line and its conjugate diameter meet the line in O and the polar of P meet the line in P′, then*

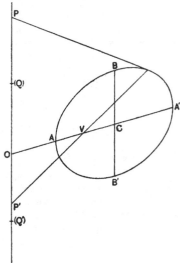

$$\frac{PQ.PQ'}{CB^2} = \frac{PO.PP'}{CB^2} = a \ constant.$$

The points P, P' are a pair of conjugate points of the involution of which Q and Q' are the double points (imaginary) and O is the centre. Hence $PQ.PQ'$ equals $PO.PP'$.

Also
$$\frac{PO.PP'}{CB^2} = \frac{PO^2 - OP.OP'}{CB^2}$$

$$= \frac{PO^2 + OA.OA'\dfrac{CB^2}{CA^2}}{CB^2} \quad \text{(Art. 38)}$$

$$= \frac{PO^2}{CB^2} + \frac{CO^2 - CA^2}{CA^2} = \frac{PO^2}{CB^2} + \frac{CO^2}{CA^2} - 1$$

$$= \text{a constant (by (i)).}$$

From this result it follows that:

(*a*) *Carnot's theorem holds, when the conic meets one or more sides of the triangle in imaginary points.*

(*b*) *Newton's theorem* (Art. [104 *a*]) *holds when the points of intersection of the chords with the conic are imaginary, also the deduction* (i) *holds in this case.* Hence *the imaginary tangents from any real point to a conic are in the ratio of the parallel semi-diameters.*

These results follow from the extension of Carnot's theorem contained in (*a*) in the same way that the corresponding results follow from Carnot's theorem.

(*c*) *If a system of conics be described through four points, a conic of the system which meets a straight line in imaginary points determines on the line a pair of conjugate points of the involution determined on the line by the conics of the system which meet it in real points.*

This follows from (*b*) by the method of Art. [101 (*b*)].

(*d*) *If three conics intersect in the same four points, the ratio of two tangents from a variable point on one conic to the two other conics is in a constant ratio to the ratio of the diameters of the two latter conics, which are parallel to the tangents.*

Consider any two points P and P' on the conic from which the tangents are drawn. Let PP' meet the other conics in Q, Q' and R, R'. Then PP', QQ', RR' are pairs of conjugate points of an involution.

Therefore
$$\frac{PQ.PQ'}{PR.PR'} = \frac{P'Q.P'Q'}{P'R.P'R'}.$$

Let d and l be the semi-diameters of the conics through Q, Q' and R, R' which are parallel to PP'.

Then
$$\frac{\dfrac{PQ \cdot PQ'}{d^2}}{\dfrac{PR \cdot PR'}{l^2}} = \frac{\dfrac{P'Q \cdot P'Q'}{d^2}}{\dfrac{P'R \cdot P'R'}{l^2}} = \text{a constant for different positions of } P.$$

Let t_1 and t_2 be tangents from P to the two conics and d_1 and d_2 the semi-diameters of these conics parallel to these tangents.

Then
$$\frac{\dfrac{PQ \cdot PQ'}{d^2}}{\dfrac{PR \cdot PR'}{l^2}} = \frac{\dfrac{t_1^2}{d_1^2}}{\dfrac{t_2^2}{d_2^2}} = \text{a constant for different positions of } P.$$

$$\therefore \frac{t_1}{t_2} \cdot \frac{d_2}{d_1} = \text{a constant for different positions of } P \text{ on the conic.}$$

43. The corresponding theorems for the parabola are as follows:

(i) *If P be a fixed point through which a variable real line PO is drawn, and the diameter of a parabola at the point A, at which the tangent is parallel to OP, meet PO at O, and S be the focus of the parabola, then* $\dfrac{PO^2 + 4SA \cdot OA}{SA}$ *is constant.*

Draw an equal parabola through the point P, having its axis in the same straight line as the given parabola. Let S' be the focus of this parabola and A' the point where OA meets this parabola.

Then
$$\frac{PO^2 + 4SA \cdot OA}{SA}$$
$$= \frac{PO^2 + 4S'A' \cdot (A'A - A'O)}{SA}$$
$$= \frac{PO^2 - 4S'A' \cdot A'O + 4S'A' \cdot A'A}{SA}$$

But by Art. [137] (1), $PO^2 - 4S'A \cdot A'O$ is zero.

Therefore $\dfrac{PO^2 + 4SA \cdot OA}{SA} = 4A'A$, which is constant.

(ii) *If P be a fixed point through which a variable line PO is drawn to meet a parabola in imaginary points Q, Q', and if A be the point of contact of the parallel tangent, S the focus, and the diameter through A meets the line PO in O and the polar of P meets it in P', then*
$$\frac{PQ \cdot PQ'}{SA} = \frac{PO \cdot PP'}{SA} = \text{a constant.}$$

The points P and P' are a pair of conjugate points of the involution of which Q and Q' are the double points (imaginary) and O is the centre. Hence $PQ \cdot PQ'$ equals $PO \cdot PP'$.

But $\quad \dfrac{PO \cdot PP'}{SA} = \dfrac{PO^2 - OP \cdot OP'}{SA} = \dfrac{PO^2 + 4SA \cdot AO}{SA}$ (Art. [137])

$$= \text{a constant by (i).}$$

Hence it follows :

(a) *That Carnot's theorem holds for a parabola when one or more sides of the triangle meet the parabola in imaginary points.*

(b) *Newton's theorem* (Art. [104(a)]) *holds for the parabola when the points of intersection of the chords with the parabola are imaginary. Hence also the tangents, real or imaginary, from any real point to a parabola are in the ratio of the square roots of the distances of their points of contact from the focus.*

44. If by means of common tangents the self-conjugate triangle of two conics can be constructed, Art. 42 (ii) renders the construction of chords of intersection of the conics possible.

Let A be a vertex of the common self-conjugate triangle and ABB' a common chord of the conics which passes through A. Let C and C' be the centres of the conics and let AC, AC' and CC' meet the conics in D, D', K and K'. Let d and d' be the semi-diameters of the conics parallel to ABB'.

Then

$$\frac{AB \cdot AB'}{d^2} = \frac{AC^2 - CD^2}{CD^2} = \frac{AC^2}{CD^2} - 1$$

and

$$\frac{AB \cdot AB'}{d'^2} = \frac{AC'^2 - C'D'^2}{C'D'^2} = \frac{AC'^2}{C'D'^2} - 1,$$

$$\therefore \quad \frac{d'^2}{d^2} = \frac{\dfrac{AC^2}{CD^2} - 1}{\dfrac{AC'^2}{C'D'^2} - 1}$$

Similarly

$$\frac{OB \cdot OB'}{d^2} = \frac{OC^2 - CK^2}{CK^2} \quad \text{and} \quad \frac{OB \cdot OB'}{d'^2} = \frac{OC'^2 - C'K'^2}{C'K'^2}$$

$$\therefore \quad \frac{\dfrac{OC^2}{CK^2} - 1}{\dfrac{OC'^2}{C'K'^2} - 1} = \frac{d'^2}{d^2} = \frac{\dfrac{AC^2}{CD^2} - 1}{\dfrac{AC'^2}{C'D'^2} - 1}$$

This relation, combined with the fact that $CO + OC' = CC'$, enables the values of CO and $C'O$ to be found.

45. *Given two straight lines a and b as the double rays real or imaginary of a real involution, to determine the two other pairs of lines a_1, b_1 and a_2, b_2 connecting their points of intersection with a conic.*

Given two points A and B as the double points real or imaginary of a real involution, to determine the two other pairs of points A_1, B_1 and A_2, B_2 in which the tangents from these points to a conic intersect.

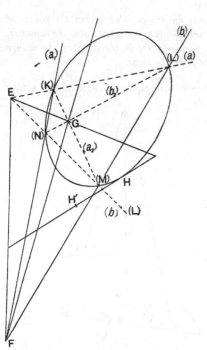
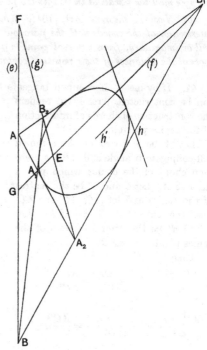

Let *a* and *b* meet the conic in *K, L, M, N*. Let *EFG* be the diagonal points triangle of this quadrangle and let *e, f, g* be its sides.

Let the tangents be *k, l, m, n*. Let *efg* be the diagonal triangle of this quadrilateral and let *E, F, G* be its vertices.

Draw *HH′* a tangent at any point *H* on the conic; every point on *HH′* is a conjugate of *H* with respect to the conic.

Draw *h* any tangent to the conic. Every line through its point of contact is a conjugate of *h* with respect to the conic.

Let the conjugate of EH with respect to a and b meet HH' in H'. Then H' is the conjugate of H with respect to another conic through K, L, M, N, viz. the lines a, b. Therefore H and H' are conjugates with respect to every conic through K, L, M, N, and therefore with respect to the lines a_1, b_1 and a_2, b_2.

Hence the following is the construction :

The line GF (e) is determined as the polar of E. EG and EF are determined as the common conjugates of the given involution and of that determined by the conic at E.

The lines a_2 and b_2 are determined as lines through G harmonic conjugates of f and e and also of the lines joining H and H' to G.

Similarly a_1 and b_1 are determined.

Let the conjugate of he with respect to A and B be joined to the point of contact of h by h'. Then h' is the conjugate of h with respect to another conic touching k, l, m, n, viz. the points A, B. Therefore h and h' are conjugates with respect to any conic touching k, l, m, n, and therefore with respect to A_1, B_1 and A_2, B_2.

Hence the following is the construction :

The point gf (E) is determined as the pole of e. eg and fe are determined as the common conjugates of the given involution and of that determined by the conic on e.

The points A_2 and B_2 are determined as points on g harmonic conjugates of F and E and of the points where h and h' meet g.

Similarly A_1 and B_1 are determined.

Particular case :

If A and B (on the right-hand side) are the circular points at infinity, the construction for A_1, B_1 and A_2, B_2 is as follows :

E, the pole of the line e (AB) now the line at infinity, becomes the centre of the conic.

f and g, by the harmonic property of the quadrilateral, are harmonic conjugates of the lines joining A and B to E and are, therefore, at right angles. f and g (since efg is self-conjugate) are conjugate lines with respect to the conic. They are therefore a pair of conjugate diameters of the conic which are at right angles, i.e. the axes of the conic.

A_1 and B_1 are harmonic conjugates of E and the point where f meets the line at infinity. They are therefore at equal distances on an axis of the conic from the centre.

h and h' are conjugate lines of an involution whose double elements are obtained by joining their point of intersection to A and B, the circular points at infinity. Therefore h and h' are at right angles and h' is the normal at the point where h touches the conic. Hence A_1 and B_1 are harmonic conjugates of the points where any tangent and normal meet the axis of the conic. They are, therefore, according to the usual definition, a pair of foci of the conic.

Hence the following definition of the foci of a conic is arrived at.

The foci of a conic are the four points real or imaginary in which the tangents from the circular points at infinity intersect.

A directrix is the polar of the corresponding focus. Hence a directrix is the chord of contact of a pair of tangents from the circular points at infinity to the conic.

Conjugate lines through a focus are at right angles. The lines joining a focus to the circular points at infinity are tangents to the curve and they are therefore the double elements of the involution on the line at infinity, determined by pairs of conjugate lines through the focus. But this must be an involution made up of pairs of lines at right angles because its double elements pass through the circular points at infinity. Therefore the conjugate lines through a focus must be at right angles.

46. The foci may also be constructed by means of a Poncelet figure.

(a) Let the curve be an ellipse. Form the Poncelet figure for the major and minor axes looking upon lines parallel to the minor axis as measured in imaginary units. The imaginary branch of the curve is a hyperbola touching the ellipse at A and B, the ends of the major axis. The tangents to these branches from the circular points at infinity are lines inclined at angles of $45°$ to the axis AB, such lines replacing in the graph those which are inclined at an angle $\tan^{-1} i$ to the axis of x. From symmetry, they must form a square with two vertices, F and F'', on the axis AB and two, F_1 and F_1', on the axis DE. Looking upon the hyperbola as real, F, F'', F_1 and F_1' lie on the director circle so that the distances CF, CF'', CF_1 and CF_1' are each equal to the square root of the difference of the squares of the semi-axes of the hyperbola. As CF and CF'' are real the points F and F'' are real. As

CF_1 and CF_1' are expressed in imaginary units, F_1 and F_1' are imaginary points. $FF_1F'F_1'$ is a semi-real square of the second kind (Art. 27).

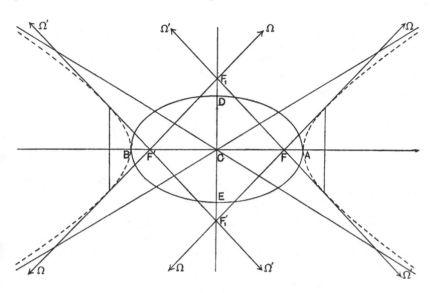

(b) Let the curve be a hyperbola. In this case the imaginary branch is the ellipse of (a). Tangents making angles of 45° with the major axis determine the foci, two on AB and two on ED. Their distances from C will be equal to the square roots of the sum of the squares of the semi-axes of the ellipse. Those on AB will be real and those on DE imaginary.

47. The existence of the foci of a conic may be explained as follows.

Through a real point A (see figure, Art. 45) a pair of conjugate imaginary lines a_1 and a_2 may under certain circumstances be drawn so as to touch a real conic.

Similarly from a second real point B a second pair of conjugate imaginary lines b_1 and b_2 may be drawn to touch the same conic.

For the real points A and B a pair of conjugate imaginary points may be substituted. In this case a_1 and a_2 and also b_1 and b_2 are no longer pairs of conjugate imaginary lines, but a_1 and b_2 are a pair of conjugate imaginary lines, as are also a_2 and b_1. Hence the points a_1b_2 and a_2b_1 (A_1 and B_1 in the figure) are real. Also the point B_2 given by a_1b_1 is the conjugate imaginary point of the point A_2 given by the lines a_2 and b_2, which are the conjugate imaginary lines of a_1 and b_1. Hence A_2 and B_2 are a pair of conjugate imaginary points. Hence the lines AB, A_1B_1 and A_2B_2 are real. Also $(A_1B_1EG)=(A_2B_2EF')=(ABGF)=-1$, and therefore E is the pole of AB.

Let GF be the line at infinity. Then E is the centre of the conic and $A_1E = EB_1$ and $B_2E = EA_2$. EG and EF being conjugate lines through E are a pair of conjugate diameters.

Let A and B be the circular points at infinity. Then EG and EF, which are harmonic conjugates of EA and EB, are at right angles. Hence, since a_1 and a_2 are parallel, as are b_1 and b_2, $A_1A_2B_1B_2$ is a semi-real square of the second kind (Art. 27), and it is seen that the foci are two real points A_1 and B_1 and a pair of conjugate imaginary points A_2 and B_2, such that $EA_1 = iEA_2$.

48. Intersections of two conics.

(1) *Two conics cannot intersect in more than four points.*

If possible let them intersect in A, B, C, D, P. Then the pencil $(P . ABCD)$ must have the same anharmonic ratio for both conics. Therefore every point on both conics subtends a pencil of the same anharmonic ratio at A, B, C, D. Therefore the conics coincide.

(2) *Every two conics with real branches have two real chords of intersection which may be coincident.*

In Art. [125] it was shown that in the case of every pair of conics there are two real lines (which may be coincident) on which the conics determine the same involution. The double points of these involutions are the four points of intersection of the conics. Therefore, excluding the special case when the chords are coincident, there are three cases for consideration :

(i) When the conics intersect in four real points,

(ii) When the conics intersect in two real and a pair of conjugate imaginary points,

(iii) When the conics intersect in two pairs of conjugate imaginary points.

(i) In this case, the common inscribed quadrangle is real and its diagonal points triangle is a real common self-conjugate triangle of the two conics.

(ii) In this case, two of the points of intersection are real and two are conjugate imaginary points. They form a semi-real quadrangle of the second kind. One vertex and the opposite side of the common self-conjugate triangle are real. The other vertices are a pair of conjugate imaginary points and the other sides a pair of conjugate imaginary lines.

(iii) In this case, the vertices of the quadrangle are two pairs of conjugate imaginary points. They form a semi-real quadrangle of the first kind. The diagonal points triangle of this quadrangle is real and is a common self-conjugate triangle of the two conics.

49. (1) *To construct graphically the imaginary points of intersection of two conics with real branches.*

(a) Let the conics intersect in a pair of real points. Then by Art. [125] the real chord joining their pair of imaginary points of intersection can be constructed. Let this chord be a. Let a' and a'' be the diameters of the conics parallel to a. Construct the diameters b and c of the conics which are conjugate respectively to a' and a'' (see figure, Art. 99).

For the conjugate diameters a' and b construct the graph of the first conic in which imaginary distances are measured parallel to a'.

Similarly for the conjugate diameters a'' and c construct the graph of the second conic in which imaginary distances are measured parallel to a''.

These two graphs intersect in the required points.

Let O be the mean point of the two conjugate imaginary points of intersection L and M. Then the imaginary coordinates of these points are measured from O along a. Since this is the case the diameters b and c must intersect on a at the point O.

(b) Let the conics intersect in two pairs of conjugate imaginary points. Then by Art. [125] a pair of common real chords of the conics can be constructed. The diameters of the two conics conjugate to the diameters parallel to these chords will intersect in pairs on the chords in question and the two pairs of graphs corresponding to these two pairs of conjugate diameters will intersect in the required points. In the two graphs imaginary lengths must be measured parallel to the two common chords.

(2) *To construct the points of intersection of a conic, having a real branch, with an imaginary straight line.*

Denote the conic by S and the imaginary straight line by l. Let l' be its conjugate imaginary line. Then l and l' are the double rays of a real overlapping involution pencil. They will intersect the conic in two pairs of conjugate imaginary points which form a semi-real quadrangle. The connectors of the pairs of conjugate imaginary points are two real straight lines. These may be constructed by means of Art. 45.

Construct the diameter of the conic parallel to either of these lines and its conjugate diameter. The Poncelet figure for the conic may be constructed for these diameters.

Construct the line parallel to this same line through the point of intersection of the conjugate imaginary lines, and its conjugate in the involution which determines the pair of conjugate imaginary lines. The Poncelet figure of the pair of imaginary lines, for these lines, may be constructed (Arts. 76 and 133). The intersections of the two figures in question give two of the four points of intersection of the conic and the pair of imaginary straight lines. The other two points may be similarly constructed.

50. From the preceding pages it will be seen that the effect of giving an interpretation to the imaginary is to do away with restrictions, which are imposed in ordinary geometry. One of these is that any two sides of a triangle must be greater than the third. A triangle may, taking the imaginary into account, have one side greater than the sum of the other two. Such a triangle may be constructed as follows.

Let a and b be the two sides the sum of which is less than the third side c.

Let A and B be the ends of c. With centre A and radius b describe a circle, and with centre B and radius a describe another circle. Let l be the radical axis of these circles. The circles intersect in two points on l which are given as the points of intersection of their hyperbolic branches described with axes parallel to AB and l. Either of these points is a vertex of the required triangle. The distances of these points from the point of intersection of AB and l are purely imaginary quantities.

EXAMPLES

(1)   If two conics have four imaginary points of intersection, show that they have a real common self-conjugate triangle and one real pair of common chords.

(2)   If A, A' be two paired elements of an elliptic involution, there is one and only one other pair which divide AA' harmonically.   Apply this to determine the imaginary line joining two given coplanar imaginary points.

(3)   Two hyperbolae have the same asymptotes: prove that they cannot intersect

(4)   Show that the four directrices of a conic are chords of intersection of the conic and its director circle.

(5)   One side of a triangle is a real line, the other two meet in a real point, and each passes through one of the circular points at infinity ; required the orthocentre.

(6)   Three given conics touch the same pair of straight lines: construct the conic which touches these lines and is such that the points of contact of any common tangent to this conic and one of the given conics are conjugate points with respect to the straight lines.

Examine the case when the straight lines pass through the critical points.

(7)   Prove that the construction of Art. 37 holds for the imaginary points o intersection of two similar and similarly situated ellipses.

Project on a parallel plane from a real point.

(8)   If a pair of conjugate imaginary lines are tangents to a real conic, their points of contact are conjugate imaginary points.

(9)   Prove the following construction for the graphs of the imaginary points of intersection of a straight line l (real) with a conic:

Let M be the pole of l.   Take M' the inverse point of M with respect to the director circle, and let l meet MM' at O.   Then the circle through M, M', whose centre is on the perpendicular to l through O, determines the required points.   (See Gaskin's theorem, Art. [138].)   Every circle through M, M' determines a pair of conjugate points on l.

(10)   In example (9) show that the circle whose centre is on the perpendicular to l at O and which cuts orthogonally the circle described on PP' as diameter, determines the points of intersection of the conic with l.

In a real plane Perspective.

[In the following the pair of conjugate imaginary points which correspond to the circular points at infinity are termed the *vanishing circles*.]

(11)   Prove that if the centre of perspective S be the centre of a circle, S is the focus of the corresponding conic and the vanishing line its directrix.

(12)   Prove that a system of rectangular hyperbolae have for their plane perspectives a system of conics which determine on the vanishing line an involution of which the vanishing circles are the double points.

(13)   Prove that the plane perspectives of a system of concentric circles form a system of conics which have double contact at the vanishing circules.

(14)   Deduce by (13) for a system of conics having double contact at conjugate imaginary points the properties which are proved in Arts. [130] and [131] for conics having double contact at real points.

(15)   Prove that a system of coaxal circles and a system of confocal conics may be looked upon as the correlatives of each other.

(16)   From (15) deduce the properties of confocal conics set forth in Art. [140] from those of coaxal circles.

(17)   Prove that the plane perspectives of a system of similar conics is a system of conics, which cut the vanishing line in constant anharmonic conjugates of the vanishing circules.

CHAPTER III

ANGLES BETWEEN IMAGINARY STRAIGHT LINES. MEASUREMENT
OF IMAGINARY ANGLES AND OF LENGTHS ON IMAGINARY
STRAIGHT LINES

51. Imaginary lines and imaginary angles.

On reference to Art. 1 it will be seen that a real line contains:

(1) An infinite number of real points determined by their real distances from some given base point.

(2) An infinite number of purely imaginary points determined by their purely imaginary distances from the base point.

(3) An infinite number of infinite systems of imaginary points, whose determining distances with reference to the base point are complex quantities. Each infinite system may be obtained by measuring purely imaginary lengths from some real point of (1) or by measuring real lengths from some purely imaginary point of (2).

The base point may be either real or imaginary. Any point, real or imaginary, on the line may be taken as this point. Points, real, purely imaginary or complex, are such with reference to the base point. If a different point, real or imaginary, be taken as base point the nature of certain of the points considered will be different. In itself however the base point is neither real nor imaginary.

The point at infinity on the line is of the same nature as the base point. It can be regarded as either real or imaginary and it belongs to the real and to the imaginary system of points. The determining distance of the base point is 0 and that of the point at infinity on the line $\frac{1}{0}$. These quantities are of the same nature. If it is conceivable to divide an infinite length into finite portions, then the length from the base point to infinity may be regarded as divided into an infinite number of real units of length and also into an infinite number of purely imaginary units of length.

Consider any pair of real lines s and s', which intersect in a real point S. Rotate the line s' round S till s and s' coincide, i.e., through an angle $\widehat{s's}$. In this way the two straight lines are made to coincide, as do also the systems of points, real and imaginary, on them.

Consider two imaginary points which do not lie on the same real straight line. By Axiom I. they have definite positions. Hence the line joining them has a definite position (see Art. 11) and the points, since they have definite positions, are at some definite distance apart. This distance generally must be a function of the real and imaginary quantities, by which the positions of the points are determined, but at present no attempt is made to define or measure this distance. It follows however that there is a measure of this distance. Hence, as along a real line, distances real and imaginary can be measured along an imaginary line. A base point can be taken, which may be the real point on the line, and from it real and purely imaginary lengths can be measured, and from the points so determined purely imaginary and real lengths may in their turn be measured. Therefore the nature of the systems of points on an imaginary line is the same as on a real line. The difference between a real and an imaginary line does not lie in the nature of the points on the lines in regard to themselves nor in the lines themselves, but in respect to the relation of the lines to other lines and to points which are not situated on the lines. In fact all lines real or imaginary have the same characteristics.

Consider two imaginary straight lines s and s' in the same plane. They intersect in a point A, which is generally imaginary but may be the real point on both lines. Consider A as the base point of systems of points on the lines s and s'. These lines s and s' have by Art. 11 definite positions.

It is now assumed that by a rotation of s' round the point A some point of s' (real with respect to A) can be brought into coincidence with some point on s (real with respect to A). If this be done the straight lines s and s' in the new position of s' must coincide, for (Art. 11) no two different straight lines real or imaginary can join the same pair of points. The measure of the amount of rotation necessary to bring the lines s and s' into coincidence is termed the angle between s and s' in their original position. The measure of this, whatever system of measurement is used, must as a general rule depend on imaginary lengths and being a function of such lengths is termed an imaginary angle. The assumption made in the preceding may be embodied in a second axiom as follows:

Axiom II. *Either of two given straight lines, real or imaginary, may be superposed on the other by a motion of rotation through a definite angle about their point of intersection* *.

---

* Hereafter it will be seen that there is an apparent exception to the principle laid down in this axiom, see Art. 78.

The rotation may be in a positive or a negative direction. After the line s' has been brought into coincidence with the line s, it may be further rotated round A so as to come into coincidence with a third line s'', which passes through A. Hence if $\widehat{ss'}$ denote the angle between the lines s and s', it follows that

$$\widehat{s's''} = \widehat{s's} + \widehat{ss''}.$$

Hence angles, real or imaginary, at a point may be measured from a base line through the point, in the same way that distances, real or imaginary, can be measured along a straight line from a base point on the line.

NOTE. (1) It does not follow that, if A is a real point and s and s' are imaginary lines, the same rotation round A which brings s' into coincidence with s will bring the conjugate imaginary line of s' into coincidence with the conjugate imaginary line of s. The angle $\widehat{s's}$ will usually be an imaginary angle and its measure will involve "i." To bring the conjugate imaginary line of s' into coincidence with the conjugate imaginary line of s the angle of rotation must be measured by a quality, which is the imaginary conjugate of the measure of $\widehat{s's}$, i.e., the rotation must be through the conjugate imaginary angle of $\widehat{s's}$, if such a term may be used. After an imaginary displacement along a straight line points which had been conjugate imaginary points cease to be so. This is also the case after a rotation through an imaginary angle.

(2) The coefficients in an analytical equation in x and y perform a double duty:

(i) They determine the dimensions of the curve and incidentally its nature.

(ii) They determine its position.

All curves, which are of the same nature and have the same dimensions, may be looked upon as the same curve. Thus all ellipses with semi-major and semi-minor axes a and b are the same curve, only displaced by a motion of translation, of say the centre, and a motion of rotation, of say the major axis. Invariants (geometrical) are of course functions of the quantities which give the dimensions of the curve.

In the equation of a real straight line the coefficients are entirely employed to determine position. All straight lines are therefore the same straight line displaced by a motion of translation of some point on the line and a motion of rotation round some point.

That this is the case with imaginary as well as with real straight lines is the assumption of Axiom II.

52. Without at present attempting to define the measure of an imaginary angle there are certain consequences of the preceding which may be noticed.

(1) The angle between two imaginary lines depends on lengths some of which as a general rule are real and some imaginary. Therefore in whatever way the measure of this angle is expressed it must be

of the form $\alpha + \alpha_i$, where α is the part of the angle which can be constructed as a real angle and α_i the part which depends on the imaginary. It is clear that no imaginary angle can equal a real angle. The angle α_i must not however be confused with the angle $i \cdot \alpha$.

(2) If the angle which an imaginary line makes with a real line be $\alpha + \alpha_i$ then the angle which its conjugate imaginary line makes with the same real straight line is $\alpha - \alpha_i$.

(3) *The internal and external bisectors of the angle between a pair of conjugate imaginary lines are real.* (See also Art. 66.)

For if the lines make angles $\alpha + \alpha_i$ and $\alpha - \alpha_i$ with any real line through their point of intersection, their bisectors make angles

$$\frac{\alpha + \alpha_i + \alpha - \alpha_i}{2} \text{ and } \frac{\alpha + \alpha_i + \alpha - \alpha_i + \pi}{2}$$

with the same straight line.

But these angles are α and $\alpha + \dfrac{\pi}{2}$ which are real.

Hence it follows that *if the angle between a real line and an imaginary line, which meets it in a real point, be expressed in the form $\alpha \pm \alpha_i$, then α is a measure of the real angle between the real line and one of the bisectors of the angle between the imaginary line and its conjugate imaginary line.* (See Art. 66.)

(4) The sum of the angles of an imaginary triangle is π.

(5) The following among other elementary theorems given in Hall and Stevens' *Geometry* hold when the lines mentioned in the enunciations are imaginary, viz., 1, 2, 3, 4, 6, 17 and 18.

Among these are the following:

(*a*) The vertical and opposite angles between lines, real or imaginary, are equal.

(*b*) The sum of the angles at any point on the same side of a line, real or imaginary, is two right angles.

(*c*) The external angle of any triangle equals the sum of the internal and opposite angles.

EXAMPLES

(1) Prove that the line $y(b+ib') - x(a+ib') = 0$ must be turned about the origin through an angle θ, where $\cot \theta = \dfrac{aa' + bb'}{ab' - ba'} - i\,\dfrac{a^2 + b^2}{ab' - ba'}$, to change it into the real line $yb - ax = 0$.

(2) Prove that the bisectors of the angle between the pair of conjugate imaginary lines $y^2 + m^2x^2 = 0$ are the pair of real lines $xy = 0$.

This follows from the fact that the bisectors of the angles between the lines $ax^2 + 2hxy + by^2 = 0$ are given by the equation $h(x^2 - y^2) = (a - b)xy$.

53. Parallel straight lines.

Consider a system of points, real and imaginary, on a real line v' in a plane σ' (cf. Art. [20], etc.). This line may be projected from a real centre S into the line at infinity in a plane σ. Through each of the real points on v' pass an infinite number of real and an infinite number of imaginary straight lines. To the real straight lines correspond in σ a system of real parallel straight lines passing through the same point at infinity. The imaginary straight lines will also pass through this point at infinity and, in view of the fact that they do not intersect the system of real lines in any points at a finite distance, they may be regarded as forming a system of lines parallel to themselves and to the real system.

Through each imaginary point on v' an infinite number of imaginary straight lines pass. These correspond to a system of imaginary straight lines in σ, which all pass through the same imaginary point at infinity. Since these straight lines do not intersect in points at a finite distance they too may be termed a system of parallel imaginary straight lines. Such a system of parallel imaginary lines intersect at an imaginary point at infinity and the one real line of the system is the line at infinity.

The angle between a pair of straight lines, real or imaginary, which meet at infinity must be infinitely small. Hence, since the sum of the angles of all triangles is equal to π, *parallel straight lines, whether real or imaginary, make equal angles with every straight line, real or imaginary, in their plane.*

The following among other elementary theorems in Hall and Stevens' *Geometry* hold, when the lines mentioned in the enunciations are imaginary, viz., 13, 14, 15, 20 and 21.

Consider a system of parallel lines, real and imaginary, which pass through a real point at infinity. (Figure, Art. 55.)

Let a real line RR' of the system meet two real lines OL and OM in R and R', OL being perpendicular to the system, and let an imaginary line of the system meet the same lines in imaginary points Q and Q'.

Then, by Art. 10, regarding S as being at infinity $\dfrac{OQ'}{OQ} = \dfrac{OR'}{OR} = \cos\theta$, where θ is the real angle between OL and OM. Hence the ratio $\dfrac{OQ'}{OQ}$ is real.

54. Perpendicular lines.

(1) *Through every imaginary point a straight line can be drawn perpendicular to a real line.*

Take any imaginary line through the imaginary point and take the real point on this line. Through it draw a perpendicular to the real line. This line is real. Join the point at infinity on it to the given imaginary point. This is the required line.

(2) *Through every real point a straight line can be drawn perpendicular to a given imaginary line.*

Join the given point to the point at infinity on the imaginary line. Let this line be l. Find the line a the harmonic conjugate of l with respect to the lines joining the given point to the circular points at infinity. Then a is the required line. (See Art. 22.)

(3) *Through every imaginary point a straight line can be drawn perpendicular to a given imaginary line.*

Draw any imaginary line through the given point. Take the real point on this line. Draw by (2) a perpendicular through it to the imaginary line. Take the point at infinity on this line—which will usually be imaginary—and join this point to the given imaginary point. This is the required line.

55. Projection of an imaginary length measured along a real line upon another real line.

Let P and Q be any two imaginary points upon a real straight line OPQ and let any other real line OL make an angle θ with OPQ.

Through P and Q draw any two imaginary lines perpendicular to OL to meet it in P' and Q'. Then $P'Q'$ is termed the projection of PQ on OL.

Take R any real point on OPQ and draw RR' perpendicular to OL to meet OL in the real point R'.

Then, by Art. 10, $\dfrac{OP'}{OP} = \dfrac{OQ'}{OQ} = \dfrac{P'Q'}{PQ} = \dfrac{OR'}{OR} = \cos\theta.$

Therefore $P'Q' = PQ\,.\,\cos\theta.$

If A, B be any two real points on a real straight line and P any imaginary point on the same straight line, the sum of the projections of AP, PB, BA, on any real line is zero.

If a real line l make an angle θ with the given line, the sum of the projections in question is

$$AP\cos\theta + PB\cos\theta + BA\cos\theta.$$

But since, Art. 3, $BA = BP + PA$, this expression is zero.

Similarly if A, B, C be any three points on a real line, the sum of the projections of AB, BC, CA on any other real line is zero.

56. *Definition of*

(1) *the measure upon a real straight line of an imaginary length along an imaginary line;*

(2) *the measure of a length along an imaginary straight line;*

(3) *the sine, cosine and tangent of the angle between a real and an imaginary straight line.*

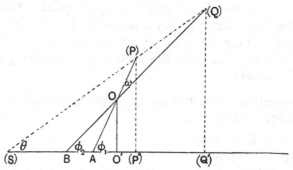

Let P and Q be any pair of imaginary points, the real lines through which, viz., OP and OQ, contain an angle ω.

Let any real straight line SBA meet PQ, OP, and OQ in S, A, and B respectively. Draw straight lines through P, Q, and O perpendicular to SBA to meet it in P', Q', O'. Of these P' and Q' are imaginary. Let ϕ_1 and ϕ_2 be the angles that PO and QO make with SBA and let θ be the angle PSP'. The angle θ is imaginary.

Then $OQ\cos\phi_2 - OP\cos\phi_1 = O'Q' - O'P' = P'Q'$ and is defined as *the measure of PQ on SBA.*

Let $P''Q''$ be the measure of PQ on a line perpendicular to SBA, then

$$OQ \sin \phi_2 - OP \sin \phi_1 = P''Q''.$$

Therefore

$$P'Q'^2 + P''Q''^2 = (OQ \cos \phi_2 - OP \cos \phi_1)^2 + (OQ \sin \phi_2 - OP \sin \phi_1)^2$$
$$= OP^2 + OQ^2 - 2 . OP . OQ (\cos \phi_2 \cos \phi_1 + \sin \phi_2 \sin \phi_1)$$
$$= OP^2 + OQ^2 - 2 . OP . OQ \cos \omega.$$

This expression, which is independent of the position of SBA, is defined as *the square of the measure of PQ*:

$\cos \theta$ is defined as *the ratio of the measure of PQ on SBA to the measure of PQ*.

$\sin \theta$ is defined as *the ratio of the measure of PQ on a line perpendicular to SBA to the measure of PQ*.

$\tan \theta$ is defined as *the ratio of $\sin \theta$ to $\cos \theta$ provided always that the measure of PQ is not zero*.

Hence*
$$\cos \theta = \frac{OQ \cos \phi_2 - OP \cos \phi_1}{\sqrt{OP^2 + OQ^2 - 2 . OP . OQ \cos \omega}},$$
$$\sin \theta = \frac{OQ \sin \phi_2 - OP \sin \phi_1}{\sqrt{OP^2 + OQ^2 - 2 . OP . OQ \cos \omega}},$$
$$\tan \theta = \frac{OQ \sin \phi_2 - OP \sin \phi_1}{OQ \cos \phi_2 - OP \cos \phi_1}.$$

It follows at once from the definition that
$$\sin^2 \theta + \cos^2 \theta = 1.$$

57. *To prove that the tangent of an angle between a real line and an imaginary line has the same value whatever imaginary points P and Q (Art. 56) are chosen to determine its value.*

Let the imaginary line be SPQ and the real line SL, and let the angle between them be θ. Let T be the real point on SPQ. Draw through T a parallel TBA to SL. Then the angle PTA equals the angle θ and it is obvious from the definition that $\tan \theta$ as obtained from the angle QTA is the same as that obtained from the angle QSL.

Let the real lines through P and Q meet TA and a line perpendicular to TA through T in A and A' and in B and B'. Let these lines intersect at O. Draw the imaginary lines PP', PP'', QQ', QQ'' through P and Q perpendicular to TA and TA'. Let the angles PAT and QBT be ϕ_1

---

* There is an ambiguity in the sign of the denominator of these expressions, which is considered hereafter.

and ϕ_2. If θ, when determined by P and T, be θ_1 then $\tan \theta_1$ is $\dfrac{PP'}{P''P}$.

If θ, when determined by Q and T, be θ_2 then $\tan \theta_2$ is $\dfrac{QQ'}{Q''Q}$.

Therefore $\tan \theta_1 = \dfrac{PA \sin \phi_1}{PA' \cos \phi_1}$ and $\tan \theta_2 = \dfrac{QB \sin \phi_2}{QB' \cos \phi_2}$.

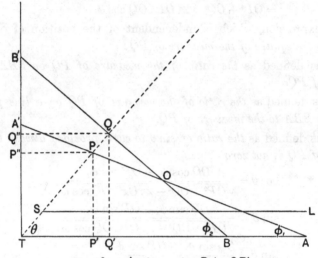

Therefore
$$\frac{\tan \theta_1}{\tan \theta_2} = \frac{\sin \phi_1 \cos \phi_2}{\sin \phi_2 \cos \phi_1} \cdot \frac{PA}{PA'} \cdot \frac{QB'}{QB}$$
$$= \frac{OB}{OA} \cdot \frac{OA'}{OB'} \cdot \frac{PA}{PA'} \cdot \frac{QB'}{QB}$$
$$= \left(\frac{BO}{B'O} : \frac{BQ}{B'Q}\right)\left(\frac{A'O}{AO} : \frac{A'P}{AP}\right)$$
$$= (BB'OQ)(A'AOP)$$
$$= \frac{(BB'OQ)}{(AA'OP)} = 1 \text{ by Art. 10.}$$

Therefore $\tan \theta_1 = \tan \theta_2 = \dfrac{PP'}{P''P} = \dfrac{QQ'}{Q''Q} = \dfrac{QQ' - PP'}{Q''Q - P''P} = $ the tangent of the angle determined by P and Q. Since, if P is fixed, Q may be any point on the line, the result follows.

Hence also $\sin \theta$ and $\cos \theta$ are independent of the positions of P and Q on the given line.

It follows from the definition (Art. 56) that

(1) If either of the lines SPQ or TBA is moved parallel to itself the values of $\sin \theta$ and $\cos \theta$ are not altered. (See Art. 53.)

(2)   If the usual convention as to the sign of a real angle is applied to imaginary angles

$$\sin(\pi - \theta) = \sin\theta, \quad \cos(\pi - \theta) = -\cos\theta.$$

(3)   If θ and θ' are the angles which an imaginary line OC makes with a real line AOB then $\sin\theta = \sin\theta'$. (See Art. 53.)

(4)   If $\theta = 0$, $\sin\theta = 0$ and $\cos\theta = 1$.

If $\theta = \dfrac{\pi}{2}$, $\sin\theta = 1$ and $\cos\theta = 0$.

Also $\sin(\theta + 2\pi) = \sin\theta$ and $\cos(\theta + 2\pi) = \cos\theta$.

(5) $\sin\left(\theta + \dfrac{\pi}{2}\right) = \cos\theta$ and $\cos\left(\theta + \dfrac{\pi}{2}\right) = -\sin\theta$.

58.   *The sum of the measures of the sides of any plane figure on a real line is zero.*

Consider any triangle ABC.   Let the real lines through A, B, C form a real triangle $A'B'C'$.   Let A'', B'', C'' be the projections of A, B, C on any real line s.   It is necessary to prove that

$$A''C'' + C''B'' + B''A'' = 0.$$

Let ϕ_1, ϕ_2, ϕ_3 be the angles, real, which the sides of $A'B'C'$ make with s.

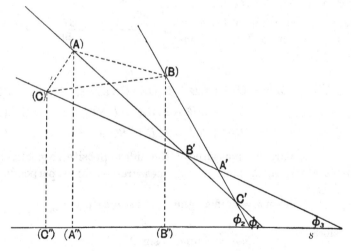

Then
$$B''C'' = A'C\cos\phi_3 - A'B\cos\phi_2,$$
$$A''C'' = B'C\cos\phi_3 - B'A\cos\phi_1,$$
$$B''A'' = C'A\cos\phi_1 - C'B\cos\phi_2.$$

Therefore $A''C'' + C''B'' + B''A''$

$= \cos \phi_1 (C'A + AB') + \cos \phi_2 (A'B + BC'') + \cos \phi_3 (B'C + CA')$

$= \cos \phi_1 (C'B') + \cos \phi_2 (A'C'') + \cos \phi_3 (B'A')$

= the sum of the projections of the sides of a real closed figure

= zero.

It should be noticed that

$$C'B' \sin \phi_1 + A'C' \sin \phi_2 + B'A' \sin \phi_3$$

is the sum of the projections of the sides of $A'B'C'$ on a perpendicular line and is therefore zero.

59. *Triangle with real lines for two of its sides.*

(1) Let A and B be two imaginary points on real straight lines CA and CB, which intersect at C at an angle ω. Let the angles CAB and CBA be α and β. Suppose also that the measure of AB is not zero and that this measure is denoted by AB.

Taking the measures of the sides of the triangle on CA,

$\quad AB \cos \alpha = CA - CB \cos \omega.$

Taking the measures of the sides of the triangle on a line perpendicular to CA,

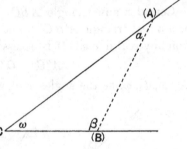

$$AB \sin \alpha = CB \sin \omega.$$

Therefore $AB^2 = CA^2 + CB^2 - 2 . CA . CB \cos \omega,$ ....(1)

also $CB^2 = CA^2 + AB^2 - 2 . CA . AB \cos \alpha.$ ....(2)

Similarly $CA^2 = BA^2 + BC^2 - 2 . BA . BC \cos \beta.$ ....(3)

(2) If a, b, c are the measures of the sides opposite respectively to the angles at A, B, C, then, taking measures on lines perpendicular to the sides,

$$a \sin \omega = c \sin \alpha \quad \text{and} \quad b \sin \omega = c \sin \beta.$$

Therefore $\dfrac{a}{\sin \alpha} = \dfrac{c}{\sin \omega} = \dfrac{b}{\sin \beta}.$

Hence in this case the measures of the sides of a triangle are proportional to the sines of the opposite angles provided none of the measures are zero.

60. *Formulae for expressing the sine and cosine of the difference—when real—of two imaginary angles in terms of their sines and cosines.*

Let an imaginary line ABT meet two real lines OA and OB, which intersect at O at an angle ω, in A and B, and let the angles OAT and OBT be ϕ_1 and ϕ_2 respectively. Then $\phi_2 - \phi_1 \equiv \omega$ (Art. 52).

Take the measures of AB and BO on AO.

$$AO = AB \cos \phi_1 + BO \cos \omega.$$

But $\quad AB = \dfrac{OA}{\sin \phi_2} \sin \omega,$

and $\quad\quad OB = \dfrac{OA}{\sin \phi_2} \sin \phi_1.$

Therefore $\quad \sin \phi_2 = \sin \omega \cos \phi_1 + \sin \phi_1 \cos \omega.$ ....(1)

Take an imaginary line $A'B'$ at right angles to AB meeting OA and OB in A' and B'. Then for ϕ_1 and ϕ_2 may be substituted $\phi_1 + \dfrac{\pi}{2}$ and $\phi_2 + \dfrac{\pi}{2}$. Hence in a similar manner

$$\sin \left(\phi_2 + \frac{\pi}{2}\right) = \sin \omega \cos \left(\phi_1 + \frac{\pi}{2}\right) + \sin \left(\phi_1 + \frac{\pi}{2}\right) \cos \omega.$$

Therefore $\quad \cos \phi_2 = - \sin \omega \sin \phi_1 + \cos \phi_1 \cos \omega.$ ....(2)

Multiply (1) by $\sin \phi_1$ and (2) by $\cos \phi_1$ and add.

Then $\quad \cos \omega = \cos (\phi_2 - \phi_1) = \cos \phi_1 \cos \phi_2 + \sin \phi_1 \sin \phi_2.$

Similarly

$$\sin \omega = \sin (\phi_2 - \phi_1) = \sin \phi_2 \cos \phi_1 - \sin \phi_1 \cos \phi_2.$$

For these results to be true it is necessary that the measures of AB and $A'B'$ should not be zero.

61. *Definition of*

(1) *The measure on an imaginary line of an imaginary length along an imaginary line and*

(2) *Of the sine, cosine and tangent of the angle between two imaginary straight lines.*

Construct a similar figure to that in Art. 56 but let the line SBA be an imaginary line.

In this case the points B, A, and O' are imaginary as are the angles ϕ_1 and ϕ_2, while θ is the imaginary angle between two imaginary straight lines.

Then as before $OQ \cos \phi_2 - OP \cos \phi_1 = O'Q' - O'P' = P'Q'$ is defined as the measure of PQ on SBA.

Let $P''Q''$ be the measure of PQ on a line perpendicular to SBA. Then
$$OQ \sin \phi_2 - OP \sin \phi_1 = P''Q''.$$
Therefore
$$
\begin{aligned}
P'Q'^2 + P''Q''^2 &= (OQ \cos \phi_2 - OP \cos \phi_1)^2 + (OQ \sin \phi_2 - OP \sin \phi_1)^2 \\
&= OP^2 + OQ^2 - 2 \,.\, OP \,.\, OQ \, (\cos \phi_2 \cos \phi_1 + \sin \phi_2 \sin \phi_1) \\
&= OP^2 + OQ^2 - 2 \,.\, OP \,.\, OQ \cos \omega \text{ (Art. 60).}
\end{aligned}
$$

This expression which is independent of the position of SBA is the same as that found in Art. 56 which has been defined as *the square of the measure of PQ.*

$\cos \theta$ is defined as *the ratio of the measure of PQ on SBA to the measure of PQ.*

$\sin \theta$ is defined as *the ratio of the measure of PQ on a line perpendicular to SBA to the measure of PQ.*

$\tan \theta$ is defined as *the ratio of sin θ to cos θ* provided always that the measure of PQ is not zero.

Hence
$$\cos \theta = \frac{OQ \cos \phi_2 - OP \cos \phi_1}{\sqrt{OP^2 + OQ^2 - 2OP \,.\, OQ \cos \omega}},$$

$$\sin \theta = \frac{OQ \sin \phi_2 - OP \sin \phi_1}{\sqrt{OP^2 + OQ^2 - 2OP \,.\, OQ \cos \omega}},$$

$$\tan \theta = \frac{OQ \sin \phi_2 - OP \sin \phi_1}{OQ \cos \phi_2 - OP \cos \phi_1}.$$

It follows at once from the definitions that
$$\sin^2 \theta + \cos^2 \theta = 1.$$

To prove that tan θ, where θ is the angle between two imaginary straight lines, is the same whatever pair of points on one of the imaginary lines is taken to determine tan θ.

This may be proved in the same way as the corresponding theorem is proved in Art. 57, the angles ϕ_1 and ϕ_2 being in this case angles between a real and an imaginary line.

Similarly $\sin \theta$ and $\cos \theta$ are independent of the positions of P and Q.

It follows from the definition that

(1) If either of the lines SPQ or TBA is moved parallel to itsel the values of $\sin \theta$ and $\cos \theta$ are not altered.   (See Art. 53.)

(2) If the usual conventions as to sign are applied to imaginary angles
$$\sin (\pi - \theta) = \sin \theta, \quad \cos (\pi - \theta) = -\cos \theta.$$

(3) If θ and θ' are the angles which an imaginary line OC makes with an imaginary line AB, then $\sin \theta = \sin \theta'$.   (See Art. 53.)

(4) If $\qquad\qquad \theta = 0, \quad \sin \theta = 0, \quad \cos \theta = 1.$

If $\qquad\qquad\qquad \theta = \dfrac{\pi}{2}, \quad \sin \theta = 1, \quad \cos \theta = 0,$

also $\qquad\qquad \sin (\theta + 2\pi) = \sin \theta, \quad \cos (\theta + 2\pi) = \cos \theta.$

(5) $\qquad\quad \sin \left(\theta + \dfrac{\pi}{2} \right) = \cos \theta, \quad \cos \left(\theta + \dfrac{\pi}{2} \right) = -\sin \theta.$

The sum of the squares of the measures of PQ on two imaginary lines at right angles gives the square of the measure of PQ.

Particular case. If the imaginary points P, Q are on the same real line, or are real, it is seen that the point O lies on this line and that *the measure on an imaginary line of a real or imaginary length PQ, which lies along a real line, is the length PQ multiplied by the cosine of the angle between PQ and the imaginary line.*

This is the same as saying that
$$\frac{SQ \cos \theta - SP \cos \theta}{PQ} = \cos \theta.$$

But SP and SQ are already defined and so is the angle.   Hence, as is necessary, an identity is arrived at.

62. *The sum of the measures of the sides of any closed plane figure, real or imaginary, on any line is zero.*

Consider the figure of Art. 58 but regard the line s as imaginary. In this case, ϕ_1, ϕ_2, ϕ_3 are imaginary angles between real and imaginary lines.

As in Art. 58 it may be shown that the sum of the projections of the sides of ABC on s is
$$\cos \phi_1 . C'B' + \cos \phi_2 . A'C' + \cos \phi_3 . B'A'. \quad \ldots\ldots\ldots\ldots(1)$$

Let S be the real point on s. Through S draw any real straight line a making an angle ϕ with s. Let ψ_1, ψ_2, ψ_3 be the angles, real, which $B'C'$, $C'A'$, $A'B'$ make with a.

Then $\qquad\qquad \phi_1 = \psi_1 - \phi, \quad \phi_2 = \psi_2 -, \phi, \quad \phi_3 = \psi_3 - \phi.$

Therefore (1) becomes

$$C'B' \cos(\psi_1 - \phi) + A'C' \cos(\psi_2 - \phi) + B'A' \cos(\psi_3 - \phi).$$

But since ϕ is the angle between a real and an imaginary line this equals by Art. 60

$$\cos\phi \left\{ C'B' \cos\psi_1 + A'C' \cos\psi_2 + B'A' \cos\psi_3 \right\}$$
$$+ \sin\phi \left\{ C'B' \sin\psi_1 + A'C' \sin\psi_2 + B'A' \sin\psi_3 \right\}.$$

But by Art. 58 both the expressions in the brackets are zero and therefore the whole expression is zero.

63. *Relations connecting the measures of the sides of an imaginary triangle and its angles.*

Let the measures of the sides of an imaginary triangle whose angles are A, B, C be a, b, c, none of the latter being zero. Taking measures of the sides of the triangle on lines perpendicular to AB and AC,

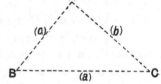

$$a \sin B = b \sin A,$$

and $\qquad\qquad\qquad a \sin C = c \sin A. \quad\dots\dots\dots\dots\dots\dots\dots(1)$

Therefore $\qquad\qquad \dfrac{a}{\sin A} = \dfrac{b}{\sin B} = \dfrac{c}{\sin C}.$

Taking the measures of the sides on a, b and c,

$$\left.\begin{aligned} a &= c \cos B + b \cos C \\ b &= a \cos C + c \cos A \\ c &= a \cos B + b \cos A \end{aligned}\right\} \quad\dots\dots\dots\dots\dots\dots(2)$$

From (1) and (2) it follows that

$$\left.\begin{aligned} c^2 &= a^2 + b^2 - 2ab \cos C \\ a^2 &= b^2 + c^2 - 2bc \cos A \\ b^2 &= a^2 + c^2 - 2ac \cos B \end{aligned}\right\} \quad\dots\dots\dots\dots\dots(3)$$

64. *To find the formulae connecting the sines and cosines of the difference of two imaginary angles with the sines and cosines of these angles.*

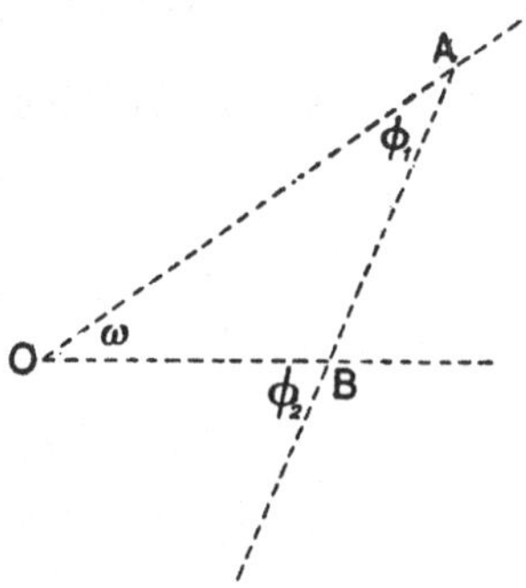

Let two imaginary straight lines OA and OB make angles ϕ_1 and ϕ_2, as in the figure, with the imaginary line AB. Then the angle AOB is $\phi_2 - \phi_1 \equiv \omega$.

Take the measures of BO and OA on BA. Then

$$AB = OA \cos \phi_1 - OB \cos \phi_2.$$

But $$\frac{AB}{\sin \omega} = \frac{OA}{\sin \phi_2} = \frac{OB}{\sin \phi_1},$$

Therefore $\sin (\phi_2 - \phi_1) = \sin \omega = \sin \phi_2 \cos \phi_1 - \sin \phi_1 \cos \phi_2$.

Increase ϕ_2 by $\frac{\pi}{2}$. Then

$$\cos (\phi_2 - \phi_1) = \cos \phi_2 \cos \phi_1 + \sin \phi_2 \sin \phi_1.$$

If ϕ_2 is taken as the internal angle at B, then

$$\omega = \pi - (\phi_1 + \phi_2) \quad \text{and} \quad AB = OA \cos \phi_1 + OB \cos \phi_2.$$

Therefore

$$\sin \omega = \sin (\phi_1 + \phi_2) = \sin \phi_2 \cos \phi_1 + \sin \phi_1 \cos \phi_2$$

and also $$\cos (\phi_1 + \phi_2) = \cos \phi_1 \cos \phi_2 - \sin \phi_2 \sin \phi_1.$$

For these formulae to hold it is necessary that the measures of the lengths from which the angles are derived should not be zero.

As these addition and subtraction formulae hold for imaginary angles, all the formulae deduced therefrom for real angles also hold for imaginary angles.

65. *To prove the general cases of Menelaus' theorem and of Ceva's theorem.*

Let the sides of an imaginary triangle ABC meet an imaginary straight line in P, Q, R as in the figure. Let the angles at P, Q, R be denoted as in the figure.

Then (Art. 63)

$$\frac{\sin \beta}{CP} = \frac{\sin \alpha}{CQ}, \quad \frac{\sin \gamma}{AQ} = \frac{\sin \beta}{AR}, \quad \frac{\sin \alpha}{BR} = \frac{\sin \gamma}{BP}.$$

Therefore

$$\sin\alpha\sin\beta\sin\gamma = \frac{CQ}{AQ}\cdot\frac{AR}{BR}\cdot\frac{BP}{CP}\sin\alpha\sin\beta\sin\gamma.$$

Hence, if none of the sines of the angles α, β, γ are zero,

$$\frac{CQ}{AQ}\cdot\frac{AR}{BR}\cdot\frac{BP}{CP}=1.$$

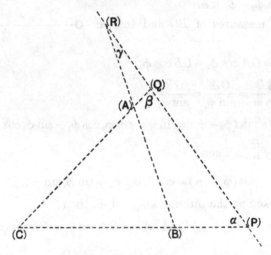

From the general case of Menelaus' theorem the general case of Ceva's theorem may be deduced as follows:

Join the vertices of any triangle ABC to any point P and let the lines AP, BP, CP meet the opposite sides in A', B', C' respectively.

Then by Menelaus' theorem for the triangle ABB',

$$\frac{AC'}{BC'}\cdot\frac{BP}{B'P}\cdot\frac{B'C}{AC}=1.$$

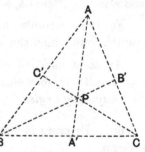

Similarly from the triangle CBB',

$$\frac{BA'}{CA'}\cdot\frac{CA}{B'A}\cdot\frac{B'P}{BP}=1.$$

Therefore

$$\frac{AC'}{BC'}\cdot\frac{BA'}{CA'}\cdot\frac{CA}{B'A}\cdot\frac{B'C}{AC}=1.$$

Therefore

$$\frac{AC'}{BC'}\cdot\frac{BA'}{CA'}\cdot\frac{CB'}{AB'}=-1.$$

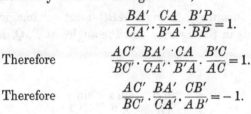

Analytical verification of the fact that the sum and difference formulae hold for a real and an imaginary angle.

Let c, $a+ib$, be the coordinates of any point P. Let $OQ=c$, $NQ=a$ and $NP=ib$. Drop PM perpendicular to ON, and MK perpendicular to OX. Let $QON=a$ and $MOP=\beta_i$, so that $NPM=a$. If

$$\sin(a+\beta_i)=\sin a \cos \beta_i + \cos a \sin \beta_i,$$

then

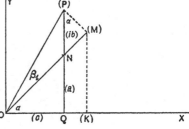

$$\frac{a+ib}{\sqrt{c^2+(a+ib)^2}}=\frac{a}{\sqrt{c^2+a^2}}\cdot\left\{\frac{\sqrt{a^2+c^2}+ib\dfrac{a}{\sqrt{a^2+c^2}}}{\sqrt{c^2+(a+ib)^2}}\right\}+\frac{c}{\sqrt{a^2+c^2}}\cdot\frac{ib\dfrac{c}{\sqrt{a^2+c^2}}}{\sqrt{c^2+(a+ib)^2}}$$

which is true.

Analytical verification of the fact that from whatever pair of lines the measure of an imaginary length is obtained its value is the same.

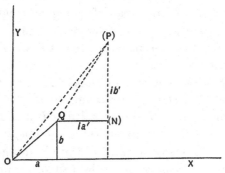

Let the coordinates of a point P referred to rectangular axes be $a+ia'$, $b+ib'$. Construct the point $Q(a, b)$ and draw QN parallel to the axis of X to meet the ordinate of P in N.

Then
$$OP^2=OQ^2+PQ^2-2OQ\cdot PQ\cos OQP$$
$$=(a^2+b^2)-(a'^2+b'^2)-2i\sqrt{a^2+b^2}\sqrt{a'^2+b'^2}\cos OQP.$$

But
$$\cos OQP=-\cos(PQN-QOX)$$
$$=-\{\cos PQN \cdot \cos QOX + \sin PQN \cdot \sin QOX\}$$
$$=-\left\{\frac{ia'}{i\sqrt{a'^2+b'^2}}\frac{a}{\sqrt{a^2+b^2}}+\frac{ib'}{i\sqrt{a'^2+b'^2}}\frac{b}{\sqrt{a^2+b^2}}\right\}$$
$$=-\frac{aa'+bb'}{\sqrt{a'^2+b'^2}\sqrt{a^2+b^2}}.$$

Therefore
$$OP^2=(a^2+b^2)-(a'^2+b'^2)+2i(aa'+bb')$$
$$=(a+ia')^2+(b+ib')^2.$$

66. *To find a real straight line such that the angle between it and a given imaginary straight line has a tangent, which is a purely imaginary quantity.*

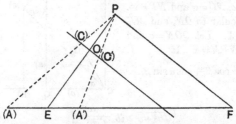

Let AP be the given imaginary straight line and P the real point on it. Take any imaginary point A on the line. Let A' be its conjugate imaginary point. Then PA' is the conjugate imaginary line of PA and the line AA' is real.

Join P to the circular points at infinity, Ω and Ω', to meet AA' in B and B'. Then B and B' are a pair of conjugate imaginary points.

Since AA', BB', are two pairs of conjugate imaginary points they have a common pair of real harmonic conjugates E and F (Art. 8). Join E and F to P. Then PE, PF are real lines.

Since PE, PF are harmonic conjugates of $P\Omega$ and $P\Omega'$ they are at right angles (Art. 22). Draw any real line COC' parallel to PF, and therefore perpendicular to PE, to meet PA, PE and PA' in C, O, C' respectively. Then C, C' are a pair of conjugate imaginary points.

Also, since $(AA'EF)$ is harmonic, $(CC'O\infty)$ is also harmonic.

Therefore O is the mean point of the pair of conjugate imaginary points C and C', and OC, OC' are purely imaginary lengths (Art. 6). Hence the tangent of the angle CPO, which is $\dfrac{OC}{PO}$, is the ratio of a purely imaginary quantity to a real quantity.

It follows from the preceding that *two conjugate imaginary lines* PA, PA' *have always a pair of real bisectors* PE, PF, *which are at right angles* (see Art. 52 (3)).

Analytical verification.

Take P as the origin and let A be $a+ib$, $c+id$, and A', $a-ib$, $c-id$.

The equation of PA is $x = \dfrac{a+ib}{c+id}\, y.$

The coordinates of the point C where a line $y = mx + k$ meets this line are

$$\frac{(a+ib)\,k}{(c+id) - m\,(a+ib)}, \qquad \frac{(c+id)\,k}{(c+id) - m\,(a+ib)}.$$

Therefore
$$PC^2 = \frac{(a+ib)^2 + (c+id)^2}{\{(c+id) - m(a+ib)\}^2} \cdot k^2.$$

This expression does not involve i if
$$m^2 + \frac{a^2 + b^2 - c^2 - d^2}{ac + bd} m - 1 = 0. \quad\text{..........................(i)}$$

Now the combined equation of PA, PA' is
$$(ay - cx)^2 + (yb - xd)^2 = 0,$$

and the equation of the bisectors of the included angles is
$$\frac{x^2 - y^2}{c^2 + d^2 - a^2 - b^2} = \frac{xy}{-(ac + bd)}. \quad\text{..........................(ii)}$$

Therefore the m's of these lines satisfy the relation
$$m^2 + \frac{a^2 + b^2 - c^2 - d^2}{ac + bd} m - 1 = 0. \quad\text{..........................(iii)}$$

This is the same equation as (i). Hence PC and PC', when CC' is parallel to a bisector of the angle AOA', have as their measures real or purely imaginary quantities. COC' is obviously perpendicular to the other bisector from the form of (ii).

67. Measurement of imaginary angles and evaluation of their sines, cosines and tangents.

Let OP be any imaginary straight line and O the real point on it.

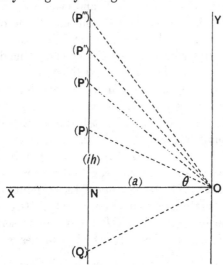

Let OQ be its conjugate imaginary line and OX and OY the bisectors (real) of the angles between these lines (Art. 66).

Take N any real point on OX at a real distance a from O. Through N draw a real line NP perpendicular to ON to meet OP in some imaginary point P.

Then as defined in Art. 56 the ratios

$$\frac{ON}{OP}, \quad \frac{NP}{OP}, \quad \text{and} \quad \frac{NP}{ON},$$

are respectively the cosine, sine and tangent of the angle XOP. Denote the angle XOP by θ_i.

Then $\qquad \cos \theta_i = \dfrac{ON}{OP}, \quad \sin \theta_i = \dfrac{NP}{OP} \quad \text{and} \quad \tan \theta_i = \dfrac{NP}{ON}.$

Now P and Q are a pair of conjugate imaginary points and N is their mean point. Therefore NP is a purely imaginary quantity, $i.h.$

Let $\dfrac{NP}{ON}$, which is the ratio of a purely imaginary quantity to a real quantity, be denoted by $i.m.$

Then $\qquad \cos \theta_i = \dfrac{1}{\sqrt{1 - m^2}}, \quad \sin \theta_i = \dfrac{i.m}{\sqrt{1 - m^2}}, \quad \tan \theta_i = i.m.$

Regard the lines OX, OY as fixed and the lines OP and OQ as being any pair of conjugate imaginary lines through O, whose bisectors are OX and OY.

As the line OP moves up to OX and eventually coincides with it, the line OQ will do the same. In the position OX the pair of conjugate imaginary lines coalesce and become the real line OX. Similarly when the line OP coincides with the line OY it also coincides with its conjugate imaginary line and becomes a real line*.

Now there is nothing inherent in a real length a and an imaginary length ih, by which it is possible to tell the relative magnitude of a compared with ih, but it is possible to tell the relative magnitudes of a series of purely imaginary quantities ih, $2ih$, $5ih$ and nih.

On the line NP take a series of lengths (all purely imaginary) with values from 0 to $i.\infty$, and let these lengths determine a series of points, $P, P', P'', \ldots$ on NP. The connectors of these points to O are lines of the system of conjugate imaginary lines of which OX and OY are the bisectors. As the imaginary line OP takes up the series of positions $OP, OP', OP'', OP''', \ldots$ the angle θ_i, which it makes with OX, will pass from a real value 0, when it coincides with OX, to a real value $\dfrac{\pi}{2}$ when it coincides with OY. The right angle XOY may therefore be divided into a series of imaginary angles in the same way in which a right angle

---

* If the line OP is given by the equation $y = imx$, the positions OX and OY correspond to the values $m = 0$ and $m = \infty$ for which values of m the line is real.

is divided into a series of equal real angles. The values of $\cos\theta_i$, $\sin\theta_i$, $\tan\theta_i$ may be evaluated for the imaginary angles into which the right angle XOY is divided.

If a real line OP is rotated round O through real angles from OX to OY, $\cos\theta$, where θ is the angle XOP, passes through all real values from 1 to 0.

If an imaginary line OP is rotated through imaginary angles—as previously set out—$\cos\theta_i$ at OX is 1. It increases as m increases, remaining real till $m = 1$, when it reaches the value ∞. This value corresponds to the critical lines through O (see Art. 22). It then becomes $i \cdot \infty$ (ignoring sign for the present) and as m increases it remains imaginary, decreasing till, when $m = \infty$ at the position OY, it is zero.

Hence, as a real line rotates from OX to OY, the cosine of the angle, which it makes with OX, takes the real values from 1 to 0 and as an imaginary line rotates from OX to OY the cosine of the angle, which it makes with OX, takes the real values from 1 to ∞ and then the imaginary values from $i\infty$ to 0 (ignoring sign for the present). Hence with the extended definition the cosine of an angle can have all values real and purely imaginary.

Similarly the sine of the real angle made by a real line with OX passes through the real values from 0 to 1. The value of $\sin\theta_i$ where θ_i is the angle made by an imaginary line—as previously described— with OX, increases through imaginary values from 0 to $i\infty$ as m increases from 0 to 1. As m increases from 1 to ∞ the value of $\sin\theta_i$ passes through real values from ∞ to 1. The cycle of possible real and purely imaginary values is thus complete.

Similarly for a real line $\tan\theta$ passes through the values 0 to ∞ as it rotates from OX to OY, and for an imaginary line $\tan\theta_i$ passes through the imaginary values from 0 to $i \cdot \infty$.

Hence with the extended definition *the sine, cosine and tangent of an angle can have all real or purely imaginary values.*

68. When an angle is written θ_i it is so written to bring to mind the fact that it is obtained by rotating an imaginary line round O and not by rotating a real line round O. The expression θ_i must not be confused with the expression $i \cdot \theta$ which is the circular measure of a real angle multiplied by the unit of imaginary length. To this expression $i \cdot \theta$ no meaning has—as yet—been attached, and as yet no measure of an imaginary angle θ_i has been defined.

When the angle θ_i takes values from 0 to $\frac{\pi}{2}$ it is necessary to find some means of measuring or constructing the different angles. The angle θ_i can be constructed from the fact that its tangent is $i\frac{h}{a}$, or $i\tan\theta$, where $\tan\theta$ is $\frac{h}{a}$. Thus the angle $30°_i$ may be represented by constructing a real angle of 30° and regarding the side opposite to the angle as measured in imaginary units. This notation will sometimes be adopted in the following pages. θ may in this case be termed the subsidiary angle of θ_i, and θ_i may be written $s\theta$, viz., the angle whose subsidiary angle is θ.

It should be noticed that while the sines and cosines of imaginary angles may be real, the tangent of an imaginary angle—as here defined —is always imaginary, and the tangent of a real angle is real. Hence in dealing with the tangent it is possible to equate real and imaginary parts regarding the tangent of an imaginary angle as imaginary.

Thus $\qquad \tan 30°_i = i\frac{1}{\sqrt{3}}, \quad \sin 30°_i = i\frac{1}{\sqrt{2}}, \quad \cos 30°_i = \frac{\sqrt{3}}{\sqrt{2}},$

$$\tan(45° + 30°_i) = \frac{1 + i\dfrac{1}{\sqrt{3}}}{1 - i\dfrac{1}{\sqrt{3}}} = \frac{1 + i\sqrt{3}}{2},$$

$$\sin(45° + 30°_i) = \frac{\sqrt{3}}{2} + i\frac{1}{2} = \frac{\sqrt{3} + i}{2},$$

$$\cos(45° + 30°_i) = \frac{\sqrt{3}}{2} - i\frac{1}{2} = \frac{\sqrt{3} - i}{2}.$$

To find $\qquad\qquad \tan^{-1}\left(\frac{1}{2} + i\frac{\sqrt{3}}{2}\right).$

Let $\qquad\qquad \tan(a + \beta_i) = \frac{1}{2} + i\frac{\sqrt{3}}{2}.$

Then $\qquad\qquad \frac{\tan a + \tan\beta_i}{1 - \tan a \tan\beta_i} = \frac{1}{2} + i\frac{\sqrt{3}}{2}.$

Let $\qquad\qquad x = \tan a, \quad y = \frac{\tan\beta_i}{i}.$

Equating real and imaginary parts

$$x = \frac{1}{2} + \frac{\sqrt{3}}{2}xy,$$

$$y = \frac{\sqrt{3}}{2} - \frac{1}{2}xy.$$

Solving these, the values $x = 1$, $y = \frac{1}{\sqrt{3}}$, and $x = -1$, $y = \sqrt{3}$ are obtained.

Taking the first solutions

$$\tan a = 1 \quad \text{therefore} \quad a = 45°,$$

$$\tan \beta_i = \frac{i}{\sqrt{3}} \quad \text{,,} \quad \beta = 30°_i.$$

Therefore

$$\tan (45° + 30°_i) = \frac{1}{2} + i \frac{\sqrt{3}}{2}.$$

Similarly

$$\tan (135° + 60°_i) = \frac{1}{2} + i \frac{\sqrt{3}}{2}.$$

To find $\tan^{-1}(a + ib)$.

In a similar manner it is found that

$$\tan a = \frac{a^2 + b^2 - 1}{2a} \pm \frac{\sqrt{(a^2 + b^2 - 1)^2 + 4a^2}}{2a},$$

and

$$\frac{\tan \beta}{i} = \frac{a^2 + b^2 - 1}{2b} \pm \frac{\sqrt{\{a^2 + (b+1)^2\}\{a^2 + (b-1)^2\}}}{2b}.$$

It should be noticed, as will be explained hereafter, that $s \cdot a + s \cdot \beta$ does not equal $s(a + \beta)$.

69. *Relations connecting sines, cosines and tangents of imaginary angles.*

From Art. 67, if $\tan \theta = \dfrac{h}{a}$,

$$\sin \theta_i = \frac{ih}{\sqrt{a^2 - h^2}} = \frac{i \tan \theta}{\sqrt{1 - \tan^2 \theta}},$$

$$\cos \theta_i = \frac{a}{\sqrt{a^2 - h^2}} = \frac{1}{\sqrt{1 - \tan^2 \theta}},$$

$$\tan \theta_i = i \frac{h}{a} = i \tan \theta,$$

where θ has all values from 0 to $\dfrac{\pi}{2}$.

Let

$$\tan \theta = \frac{e^y - e^{-y}}{e^y + e^{-y}}.$$

Then

$$y = \tfrac{1}{2} \operatorname{Log} \frac{1 + \tan \theta}{1 - \tan \theta}.$$

Substituting in the above equation in terms of y

$$\left. \begin{aligned} \sin \theta_i &= i \frac{e^y - e^{-y}}{2} = i \sinh y \\[2mm] \cos \theta_i &= \frac{e^y + e^{-y}}{2} = \cosh y \\[2mm] \tan \theta_i &= i \frac{e^y - e^{-y}}{e^y + e^{-y}} = i \tanh y \end{aligned} \right\} \quad \dots\dots\dots\dots(1)$$

Hence

$$\sin \theta_i = i \left\{ \tfrac{1}{2} \operatorname{Log} \frac{1 + \tan \theta}{1 - \tan \theta} + \frac{1}{\lfloor 3} \left(\tfrac{1}{2} \operatorname{Log} \frac{1 + \tan \theta}{1 - \tan \theta} \right)^3 \right.$$

$$\left. + \frac{1}{\lfloor 5} \left(\tfrac{1}{2} \operatorname{Log} \frac{1 + \tan \theta}{1 - \tan \theta} \right)^5 + \dots \right\}$$

and

$$\cos \theta_i = 1 + \frac{1}{\lfloor 2} \left(\tfrac{1}{2} \operatorname{Log} \frac{1 + \tan \theta}{1 - \tan \theta} \right)^2 + \tfrac{1}{4} \left(\tfrac{1}{2} \operatorname{Log} \frac{1 + \tan \theta}{1 - \tan \theta} \right)^4 + \dots$$

From equations (1) it follows that θ_i is a function of y. Let $\theta_i = f(y)$.

Then

$$\sin f(y) = i \sinh y,$$
$$\cos f(y) = \cosh y,$$
$$\tan f(y) = i \tanh y.$$

Hence $\sin f(y + z) = i \sinh (y + z)$

$$= i \{\sinh y \cosh z + \cosh y \sinh z\}$$
$$= \sin f(y) \cos f(z) + \cos f(y) \sin f(z).$$

Similar results hold for $\cos f(y)$ and $\tan f(y)$.

Hence, *when y is looked upon as the parameter, imaginary angles may be added according to the usual formulae.*

Subsidiary angles.

Let

$$\tan \theta_i = i \tanh y \quad \text{and} \quad \tan \phi_i = i \tanh z,$$

so that

$$y = \tfrac{1}{2} \operatorname{Log} \frac{1 + \tan \theta}{1 - \tan \theta} \quad \text{and} \quad z = \tfrac{1}{2} \operatorname{Log} \frac{1 + \tan \phi}{1 - \tan \phi}$$

Then

$$\sin (\theta_i + \phi_i) = \sin \theta_i \cos \phi_i + \cos \theta_i \sin \phi_i$$
$$= i (\sinh y \cosh z + \cosh y \sinh z)$$
$$= i \sinh (y + z)$$
$$= \sin \psi_i, \text{ where } y + z = \tfrac{1}{2} \operatorname{Log} \frac{1 + \tan \psi}{1 - \tan \psi}.$$

The relation between θ, ϕ and ψ is consequently given by

$$\tfrac{1}{2} \operatorname{Log} \frac{1 + \tan \psi}{1 - \tan \psi} = y + z = \tfrac{1}{2} \operatorname{Log} \frac{1 + \tan \theta}{1 - \tan \theta} + \tfrac{1}{2} \operatorname{Log} \frac{1 + \tan \phi}{1 - \tan \phi}.$$

Therefore

$$\frac{1 + \tan \psi}{1 - \tan \psi} = \frac{1 + \tan \theta}{1 - \tan \theta} \cdot \frac{1 + \tan \phi}{1 - \tan \phi}.$$

Therefore

$$\tan \psi = \frac{\tan \theta + \tan \phi}{1 + \tan \theta \tan \phi}.$$

This gives the relation connecting the subsidiary angles. The relation connecting the subsidiary angles can be obtained in a similar manner from the addition formulae for the cosine or tangent.

Hence if the sine, cosine or tangent of an imaginary angle is required it is possible to proceed in either of two ways.

Suppose that $\tan (30°_i + 60°_i)$ is required.

(a) $\tan (30°_i + 60°_i) = \dfrac{\tan 30°_i + \tan 60°_i}{1 - \tan 30°_i \tan 60°_i}$

$$= \dfrac{i \dfrac{1}{\sqrt{3}} + i \sqrt{3}}{1 - i \dfrac{1}{\sqrt{3}} i \sqrt{3}} = i \dfrac{2}{\sqrt{3}}.$$

(b) $\tan \psi = \dfrac{\tan \theta + \tan \phi}{1 + \tan \theta \tan \phi} = \dfrac{\tan 30° + \tan 60°}{1 + \tan 30° \tan 60°}$

$$= \dfrac{\sqrt{3} + \dfrac{1}{\sqrt{3}}}{1 + 1} = \dfrac{2}{\sqrt{3}}$$

Therefore $\tan (30°_i + 60°_i) = i \tan \psi = i \dfrac{2}{\sqrt{3}}$.

If $\theta_i + \phi_i = \psi_i,$

since $\tan \psi = \dfrac{\tan \theta + \tan \phi}{1 + \tan \theta \tan \phi},$

therefore $\sin \psi = \dfrac{\sin (\theta + \phi)}{\sqrt{\sin^2 (\theta + \phi) + \cos^2 (\theta - \phi)}},$

and $\cos \psi = \dfrac{\cos (\theta - \phi)}{\sqrt{\sin^2 (\theta + \phi) + \cos^2 (\theta - \phi)}}.$

The advantage of using the subsidiary angle θ as the parameter lies in the fact that (ignoring sign for the present) while it passes through all real values from 0 to $\dfrac{\pi}{2}$, $\sin \theta$ and $\sin \theta_i$ pass together through all real and purely imaginary values, while $\cos \theta$ and $\cos \theta_i$ and also $\tan \theta$ and $\tan \theta_i$ do the same. There are however advantages, as will be shown later (Art. 70), in taking y or rather iy as the measure of the imaginary angle θ_i.

According to the definition of Art. 56, if θ'_i be the angle OPN, then

$$\sin \theta'_i = \frac{ON}{PO} = \frac{1}{\sqrt{1 - m^2}}, \quad \cos \theta'_i = \frac{PN}{PO} = \frac{im}{\sqrt{1 - m^2}}$$

and $$\tan \theta'_i = \frac{ON}{PN} = \frac{1}{im}.$$

Hence

$$\sin (\theta_i + \theta'_i) = \sin \theta_i \cos \theta'_i + \cos \theta_i \sin \theta'_i$$

$$= \frac{im}{\sqrt{1 - m^2}} \frac{im}{\sqrt{1 - m^2}} + \frac{1}{\sqrt{1 - m^2}} \frac{1}{\sqrt{1 - m^2}} = 1 = \sin \frac{\pi}{2}$$

and

$$\cos (\theta_i + \theta'_i) = \cos \theta_i \cos \theta'_i - \sin \theta_i \sin \theta'_i$$

$$= \frac{1}{\sqrt{1 - m^2}} \frac{im}{\sqrt{1 - m^2}} - \frac{im}{\sqrt{1 - m^2}} \frac{1}{\sqrt{1 - m^2}} = 0 = \cos \frac{\pi}{2}.$$

This confirms the fact that $\theta_i + \theta'_i = \frac{\pi}{2}$.

The trigonometrical ratios of a purely imaginary angle may be obtained from any triangle, with two sides at right angles, on which sides the vertices are at distances, the one purely imaginary and the other real.

Generally, the sines, cosines, and tangents of a complex angle as defined in Arts. 56 and 61 are complex quantities. A complex angle can however be expressed as the sum or difference of a real and a purely imaginary angle, and therefore its trigonometrical functions can be expressed as functions of the trigonometrical functions of real and of purely imaginary angles.

70. Measure of an imaginary angle.

In Arts. 56, 60, 61 and 64 the sine, cosine and tangent of an imaginary angle were defined and it was shown that the addition and subtraction formulae, which are true for the sines and cosines of real angles, also hold for the sines and cosines of such imaginary angles.

No measure of such imaginary angles however was introduced. This leaves the subject in the same position as that, in which the trigonometry of real angles might be supposed to be, before a radian or a degree was defined.

In Art. 67 it was shown that the sine, cosine and tangent of a purely imaginary angle termed θ_i might be constructed by means of a subsidiary angle θ. Still no way of measuring the imaginary angle θ_i was laid down.

In Art. 69 it was proved that if

$$\tan \theta = \frac{e^y - e^{-y}}{e^y + e^{-y}} \quad \text{or} \quad y = \tfrac{1}{2} \operatorname{Log} \frac{1 + \tan \theta}{1 - \tan \theta},$$

then

$$\sin \theta_i = i \frac{e^y - e^{-y}}{2} = \frac{e^{-y} - e^y}{2i} = i \sinh y,$$

$$\cos \theta_i = \frac{e^y + e^{-y}}{2} = \frac{e^{-y} + e^y}{2} = \cosh y,$$

$$\tan \theta_i = i \frac{e^y - e^{-y}}{e^y + e^{-y}} = \frac{e^{-y} - e^y}{i \{e^{-y} + e^y\}} = i \tanh y.$$

Hence, since the trigonometrical functions of an imaginary angle θ_i are functions of y, it was possible to write them as $\sin f(y)$, $\cos f(y)$, $\tan f(y)$.

It was then proved that in this form the y follows the addition and subtraction formulae which hold for the sines and cosines of real angles.

In this case

$$\sin f(y) = \frac{e^{-y} - e^{+y}}{2i}, \quad \cos f(y) = \frac{e^{-y} + e^y}{2}, \quad \tan f(y) = \frac{e^{-y} - e^y}{i\{e^{-y} + e^y\}} \quad ....(1)$$

Now according to the usual theory, if y is real,

$$\sin y = \frac{e^{iy} - e^{-iy}}{2i}, \quad \cos y = \frac{e^{iy} + e^{-iy}}{2}, \quad \tan y = \frac{e^{iy} - e^{-iy}}{i\{e^{iy} + e^{-iy}\}}. \quad ...(2)$$

And if iy be an imaginary angle

$$\sin iy = \frac{e^{-y} - e^y}{2i}, \quad \cos iy = \frac{e^{-y} + e^y}{2}, \quad \tan iy = \frac{e^{-y} - e^y}{i\{e^{-y} + e^y\}}, \quad ...(3)$$

by the usual series definitions.

Comparing (1) and (3) it is seen that they are in agreement if $f(y) = iy$.

Hence the measure of an imaginary angle may now be defined as being iy, so that

$$\sin iy = \frac{e^{-y} - e^{+y}}{2i}, \quad \cos iy = \frac{e^{-y} + e^y}{2}, \quad \tan iy = \frac{e^{-y} - e^y}{i\{e^{-y} + e^y\}}$$

The trigonometrical functions of such an imaginary angle may be constructed by finding the real subsidiary angle which is given by

$$\tan \theta = \frac{e^y - e^{-y}}{e^y + e^{-y}}, \quad \text{where } y = \tfrac{1}{2} \operatorname{Log} \frac{1 + \tan \theta}{1 - \tan \theta}.$$

Then

$$\sin iy = \frac{i \tan \theta}{\sqrt{1 - \tan^2 \theta}}, \quad \cos iy = \frac{1}{\sqrt{1 - \tan^2 \theta}}, \quad \tan iy = i \tan \theta.$$

These trigonometrical functions may be graphically constructed by means of Art. 67 and the values of the sine, cosine and tangent of an imaginary angle iy may be obtained by means of the real angle θ.

For the trigonometrical functions of an imaginary angle so defined the addition and subtraction formulae hold and—as a general rule—the expression iy may be treated as the product of a real and an imaginary quantity.

It follows that a complex angle $\alpha + i\beta$ has a period 2π.

Use of meridional tables to ascertain the subsidiary angle of a given imaginary angle.

Given an imaginary angle iy the relation to find θ is

$$y = \tfrac{1}{2} \operatorname{Log} \frac{1 + \tan \theta}{1 - \tan \theta}.$$

The tables of meridional parts give v where

$$v = \frac{10,800}{\pi} \phi' = \frac{10,800}{\pi} \operatorname{Log} \frac{1 + \tan \tfrac{1}{2} \theta}{1 - \tan \tfrac{1}{2} \theta}.$$

Hence if y is given in minutes, the number $2y$ must be looked out in the table and the angle θ corresponding to the value of $2y$ can be found. One-half of this angle is the required subsidiary angle.

The values of θ in the tables are from $0°$ to $90°$. Hence the values of the subsidiary angles are for values from $0°$ to $45°$.

Summary.

These results may be summarised as follows:

The trigonometrical functions of an imaginary angle θ_i as defined in Art. 67 are

$$\sin \theta_i = \frac{im}{\sqrt{1 - m^2}}, \quad \cos \theta_i = \frac{1}{\sqrt{1 - m^2}}, \quad \tan \theta_i = im,$$

where $m = \tan \theta$. Let $\tan \theta = \dfrac{e^y - e^{-y}}{e^y + e^{-y}}$ or $y = \tfrac{1}{2} \operatorname{Log} \dfrac{1 + \tan \theta}{1 - \tan \theta}$.

The values of y for corresponding values of θ can be found from the table of meridional parts.

Then

$$\sin \theta_i = \frac{e^{-y} - e^y}{2i}, \quad \cos \theta_i = \frac{e^{-y} + e^y}{2}, \quad \tan \theta_i = \frac{e^{-y} - e^y}{i\{e^{-y} + e^y\}}.$$

By the series definition

$$\sin iy = \frac{e^{-y} - e^y}{2i}, \quad \cos iy = \frac{e^{-y} + e^y}{2}, \quad \tan iy = \frac{e^{-y} - e^y}{i\{e^{-y} + e^y\}}.$$

Hence $\theta_i = iy$, where $\tan\theta = \dfrac{e^y - e^{-y}}{e^y + e^{-y}}$ or $y = \frac{1}{2}\,\mathrm{Log}\,\dfrac{1 + \tan\theta}{1 - \tan\theta}$.

Therefore

$$\sin iy = \frac{\cdot im}{\sqrt{1 - m^2}}, \quad \cos iy = \frac{1}{\sqrt{1 - m^2}}, \quad \tan iy = im,$$

where

$$m = \frac{e^y - e^{-y}}{e^y + e^{-y}},$$

and $m = \tan\theta$, θ being the real subsidiary angle of θ_i or iy. Hence the values of $\sin iy$, $\cos iy$ and $\tan iy$ can be graphically obtained. If θ_i be written $s\,.\,\theta$, i.e., the angle whose subsidiary angle is θ, then $s\theta$ is iy,

where

$$y = \frac{1}{2}\,\mathrm{Log}\,\frac{1 + \tan\theta}{1 - \tan\theta}, \quad \text{or} \quad s\theta = i\,\frac{1}{2}\,\mathrm{Log}\,\frac{1 + \tan\theta}{1 - \tan\theta}.$$

The addition and subtraction theorems have been proved (Art. 69) to hold for y and they therefore hold for iy.

71. Values of the trigonometrical functions of an imaginary angle.

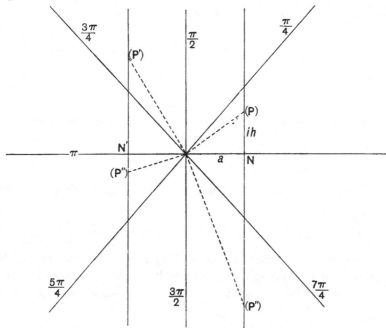

The values of the sine, cosine and tangent of iy_θ an imaginary angle have been defined as

$$\sin iy_\theta = \frac{ih}{\sqrt{a^2 - h^2}}, \quad \cos iy_\theta = \frac{a}{\sqrt{a^2 - h^2}}, \quad \tan iy_\theta = i\,\frac{h}{a}, \quad \text{where } \tan\theta = \frac{h}{a}.$$

If $h > a$ these may be written

$$\sin iy_\theta = \frac{h}{\sqrt{h^2 - a^2}}, \quad \cos iy_\theta = \frac{-ia}{\sqrt{h^2 - a^2}}, \quad \tan iy_\theta = i\frac{h}{a}$$

There is an ambiguity of sign in regard to the square root, but it will be convenient to take the above as the definitions. In order likewise to complete the cycle of values it is convenient to take a second line perpendicular to the axis of x at a distance $-a$ from the origin and to consider intercepts made by the variable line on this line, as well as on the line $x - a = 0$.

The following will then be found to be the values of the sine, cosine and tangent of the imaginary angle in the different semi-quadrants:

θ	$\sin iy$	$\cos iy$	$\tan iy$	iy
$-\frac{\pi}{4}$ to 0	$-i.\infty$ to $-i.0$	∞ to 1	$-i$ to $-i.0$	$2\pi - i.\infty$ to 2π
0 to $\frac{\pi}{4}$	$i.0$ to $i.\infty$	1 to ∞	$i.0$ to i	0 to $i.\infty$
$\frac{\pi}{4}$ to $\frac{\pi}{2}$	∞ to 1	$-i.\infty$ to $-i.0$	i to $i.\infty$	$\frac{\pi}{2} + i.\infty$ to $\frac{\pi}{2}$
$\frac{\pi}{2}$ to $\frac{3\pi}{4}$	1 to ∞	$i.0$ to $i.\infty$	$-i.\infty$ to $-i$	$\frac{\pi}{2}$ to $\frac{\pi}{2} - i.\infty$
$\frac{3\pi}{4}$ to π	$i.\infty$ to $i.0$	$-\infty$ to -1	$-i$ to $-i.0$	$\pi - i.\infty$ to π
π to $\frac{5\pi}{4}$	$-i.0$ to $-i.\infty$	-1 to $-\infty$	$i.0$ to i	π to $\pi + i.\infty$
$\frac{5\pi}{4}$ to $\frac{3\pi}{2}$	$-\infty$ to -1	$i.\infty$ to $i.0$	i to $i.\infty$	$\frac{3\pi}{2} + i.\infty$ to $\frac{3\pi}{2}$
$\frac{3\pi}{2}$ to $\frac{7\pi}{4}$	-1 to $-\infty$	$-i.0$ to $-i.\infty$	$-i.\infty$ to $-i$	$\frac{3\pi}{2}$ to $\frac{3\pi}{2} - i.\infty$

Hence the sine, cosine and tangent of any angle, real or purely imaginary, can have any value, real or purely imaginary. There is a discontinuity in the value of iy at the critical lines, where the value of iy is increased by $\frac{\pi}{2}$.

In the preceding by the introduction of the line $x + a = 0$ values of $\sin \theta_i$ and $\cos \theta_i$ have been obtained with different signs. Thus both the signs which may be given to $\sqrt{h^2 - a^2}$ have been taken into account.

Values of an imaginary angle.

Let y_θ' be the general value of $\frac{1}{2}\operatorname{Log}\dfrac{1+\tan\theta}{1-\tan\theta}$ and let y_θ be its principal value. Then since

$$\operatorname{Log}\frac{1+\tan\theta}{1-\tan\theta}=2\pi ni+\log\frac{1+\tan\theta}{1-\tan\theta},$$

it is seen that

$$iy_\theta'=-n\pi+iy_\theta. \quad\text{......................................(1)}$$

To derive the values of $\sin\theta_i$ and $\cos\theta_i$ between $\dfrac{3\pi}{4}$ and $\dfrac{7\pi}{4}$ from those between $-\dfrac{\pi}{4}$ and $\dfrac{3\pi}{4}$, iy_θ must be increased by π.

Expressing for shortness the fact that $iy_{\theta'}$ and iy_θ give the same values of the sine, cosine and tangent of θ and θ', as $iy_{\theta'}=iy_\theta$, the following results are arrived at.

(a)
$$y_{-\theta}=\tfrac{1}{2}\operatorname{Log}\frac{1-\tan\theta}{1+\tan\theta}=-\tfrac{1}{2}\operatorname{Log}\frac{1+\tan\theta}{1-\tan\theta}.$$

Therefore
$$iy_{-\theta}=-iy_\theta. \quad\text{......................................(i)}$$

On reference to the values of the sine, cosine and tangent this is seen to be true.

(b)
$$y_{\theta+\frac{1}{2}\pi}=\tfrac{1}{2}\operatorname{Log}\frac{1-\cot\theta}{1+\cot\theta}=-\tfrac{1}{2}\operatorname{Log}\frac{\tan\theta+1}{\tan\theta-1}=\pm\operatorname{Log}i-\tfrac{1}{2}\operatorname{Log}\frac{1+\tan\theta}{1-\tan\theta}$$

$$=\pm i\frac{\pi}{2}-y_\theta.$$

Therefore
$$iy_{\theta+\frac{1}{2}\pi}=\mp\frac{\pi}{2}-iy_\theta.$$

There is an ambiguity as to the sign to be taken in $\pm\dfrac{\pi}{2}$.

Let the first quadrant be that from $-\dfrac{\pi}{4}$ to $\dfrac{\pi}{4}$, the second that from $\dfrac{\pi}{4}$ to $\dfrac{3\pi}{4}$, the third that from $\dfrac{3\pi}{4}$ to $\dfrac{5\pi}{4}$ and so on. Let m denote the quadrant in which the arm of the angle θ lies. Then if the above result be written in the form

$$iy_{\theta+\frac{1}{2}\pi}=(-1)^{m+1}\frac{\pi}{2}-iy_\theta*\quad\text{..............................(ii)}$$

it will be found to be true on reference to the table of values.

Hence it follows that

$$iy_{\theta+\pi}=-(-1)^{m+1}\frac{\pi}{2}-iy_{\theta+\frac{1}{2}\pi}.$$

Hence
$$iy_{\theta+\pi}=-(-1)^{m+1}\pi+iy_\theta$$

which is in agreement with (1). If π be added or subtracted the results differ by 2π, hence this result may be stated in the form $iy_{\theta+\pi}=\pi+iy_\theta$.

---

* This result may also be written in the form

$$iy_{\theta+\frac{1}{2}\pi}=\frac{2m-1}{2}\pi-iy_\theta.$$

(c) $$y_{\frac{1}{2}\pi - \theta} = \tfrac{1}{2}\operatorname{Log}\frac{1+\cot\theta}{1-\cot\theta} = \pm\operatorname{Log} i + \tfrac{1}{2}\operatorname{Log}\frac{1+\tan\theta}{1-\tan\theta}$$

$$= \pm i\,\frac{\pi}{2} + \tfrac{1}{2}\operatorname{Log}\frac{1+\tan\theta}{1-\tan\theta}.$$

If m be defined as in (b) this formula may be taken in the form

$$iy_{\frac{1}{2}\pi - \theta} = (-1)^{m+1}\frac{\pi}{2} + iy_\theta . \quad\dots\dots\dots\dots\dots\dots\dots(\text{iii})$$

On reference to the table of values this is found to be true.

If θ be increased by $\dfrac{\pi}{2}$ (iii) becomes $iy_{-\theta} = -(-1)^{m+1}\dfrac{\pi}{2} + iy_{\theta + \frac{1}{2}\pi}$.

This, remembering that $iy_{-\theta} = -iy_\theta$, is (ii).

(d) $$y_{\theta+\pi} = \tfrac{1}{2}\operatorname{Log}\frac{1+\tan(\theta+\pi)}{1-\tan(\theta+\pi)} = \tfrac{1}{2}\operatorname{Log}\frac{1+\tan\theta}{1-\tan\theta}.$$

Hence it would seem that $iy_{\theta+\pi} = iy_\theta$. This on reference to our table of values of sine, cosine and tangent is not true. In fact our geometrical restrictions require us to assume that $\operatorname{Log}\dfrac{1+\tan(\theta+\pi)}{1-\tan(\theta+\pi)}$ has its general value and this in fact is $y_\theta - i\pi$. This result is then in agreement with those already obtained for

$$iy_{\theta+\pi} = \pi + iy_\theta.$$

72. *Construction of the sine, cosine and tangent of a purely imaginary angle by means of the imaginary branch of a circle.*

Let OX and OY be a pair of orthogonal lines. Take a circle centre O and radius a. Draw its real branch and also the two imaginary branches* $(1, i)$ and $(i, 1)$ corresponding to the axes OX and OY.

(1) Through O draw any real line OP to make a real angle with OX. Then if this line meets the real branch in P,

$$\tan\theta = \frac{PN}{ON}, \quad \sin\theta = \frac{PN}{a}, \quad \cos\theta = \frac{ON}{a}.$$

(2) Through O draw an imaginary line OP' to make an imaginary angle θ_i with OX. Let OP' meet the imaginary branch $(1, i)$ in P'. Then, since the measure of the distance of every imaginary point on the curve from O is a,

$$\tan\theta_i = \frac{P'N'}{ON'}, \quad \sin\theta_i = \frac{P'N'}{a}, \quad \cos\theta_i = \frac{ON'}{a},$$

where N' is the foot of the perpendicular from P' on OX.

---

* These are respectively the branches for which the coordinates are x, iy, and ix, y.

(3)   Through O draw an imaginary line OP'' to meet the $(i, 1)$ branch in P''.   Let this line make an angle θ_i with OX.   Then as in the previous case

$$\tan \theta_i = \frac{P''N''}{ON''}, \quad \sin \theta_i = \frac{P''N''}{a}, \quad \cos \theta_i = \frac{ON''}{a},$$

where N'' is the foot of the perpendicular from P'' on OX.   In this case $P''N'$ is real and ON'' is imaginary.

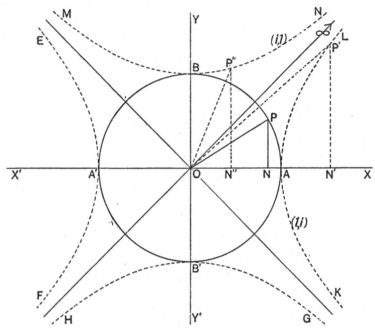

Hence, if the point P be supposed to move along the real branch from A to B, and from B to ∞ (a critical point) along the $(i, 1)$ branch, and then from ∞ to A along the $(1, i)$ branch, it will describe a closed curve in the quadrant OX, OY and the angle, real or purely imaginary, which OP makes with the axis OX is such that its tangent is $\frac{PN}{ON}$, its sine, $\frac{PN}{a}$ and its cosine, $\frac{ON}{a}$, where a is the radius of the circle, PN the ordinate and ON the abscissa of P.

In this case the trigonometrical functions are obtained from a right-angled triangle, the hypothenuse of which is a real quantity a, while one side, that measured along the axis of X, takes all real and purely imaginary values.

The values of the sine, cosine and tangent of the purely imaginary angles obtained by rotating the imaginary line through 2π round O for the different semi-quadrants are the same values as those in Art. 71, but they occur in different order.   The reason of this is as follows.

In the preceding the intersections of the line $iy = mx$ with the circle $x^2 + y^2 = a^2$ are sought for and it is found that $x^2 = \dfrac{a^2}{1-m^2}$.   If $m < 1$ the values of x are real.   This is the case along the branch AB.   If $m > 1$ there is no real value of x.   In this case the intersections of the line with the circle are given by writing the equation of the line in the form $-y = mix$.   This is a line in the quadrants 2 and 4 and it meets the circle, where $y^2 = \dfrac{a^2 m^2}{m^2 - 1}$, i.e., for real values of y, where $m > 1$.   Hence the line after having met the curve along the branch K to L meets it along the branch M to N.   Similarly it again meets the curve from E to F and finally from G to H.   If the values obtained are interchanged in such a way as to take this fact into account they are found to be in agreement with those previously given.

If, in accordance with the suggestion thrown out in Art. 126, the iy axis be taken in the direction OY' and not in the direction OY, the line $\dfrac{iy}{x} = m$ meets the circle in consecutive points, which lie from A to K, G to H, F to E, M to N, and from L to A.

73.   *Analytical verification of the fact that the definition of an imaginary angle is in agreement with the analytical theory.*

If $ax^2 + 2hxy + by^2 = 0$ be the equation of a pair of straight lines referred to rectangular axes and θ be the angle between them,

$$\tan \theta = \frac{2\sqrt{h^2 - ab}}{a+b} \quad \dotfill (1)$$

Consider the pair of conjugate imaginary lines $y - imx = 0$ and $y + imx = 0$.   Their combined equation is $y^2 + m^2 x^2 = 0$.

Substituting in (1) it is seen that

$$\tan \theta = \frac{2\sqrt{-m^2}}{1+m^2} = \frac{2im}{1+m^2}. \quad \dotfill (2)$$

If $\frac{1}{2}\theta_i$ be taken as the angle between one of them and a bisector of the angle which they contain, then

$$\tan \tfrac{1}{2}\theta_i = im.$$

Therefore

$$\tan \theta_i = \frac{2im}{1+m^2},$$

which is in agreement with (2).

Hence the angle given by the definition of Art. 67 is the same as that given by the ordinary analytical formula.

74. Eccentric angle of an ellipse.

(*a*) Construct the real branch, the $(1, i)$ branch and the $(i, 1)$ branch. Construct the auxiliary circle and its $(1, i)$ and $(i, 1)$ branches. Let a and b be the semi-major and semi-minor axes of the ellipse so that its equation is $\dfrac{x^2}{a^2} + \dfrac{y^2}{b^2} = 1$ and that of the auxiliary circle $x^2 + y^2 = a^2$. Take P_1, P_2, P_3 points on the real branch, the $(1, i)$ branch and the $(i, 1)$ branch respectively of the ellipse and let the ordinates P_1N_1, P_2N_2, P_3N_3 at these points meet the corresponding branches of the auxiliary circle in P_1', P_2', P_3' respectively. Then, if the angles $P_1'OX$, $P_2'OX$, $P_3'OX$ be θ_1, θ_2 and θ_3; θ_1 is real, θ_2 is purely imaginary and $< \tan^{-1} i$, and θ_3 is purely imaginary and $> \tan^{-1} i$, and the lengths OP_1', OP_2', OP_3' are real and equal to a.

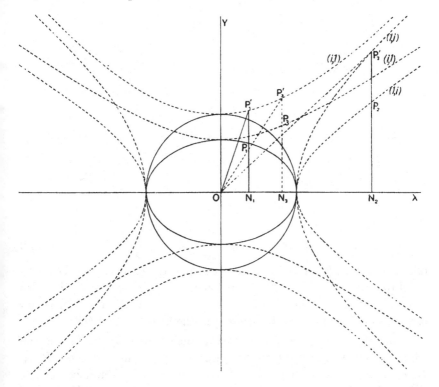

Then ON_1, ON_2, ON_3 are respectively $a \cos \theta_1$, $a \cos \theta_2$, $a \cos \theta_3$. Hence from the equation to the ellipse P_1N_1, P_2N_2, P_3N_3 are respectively $b \sin \theta_1$, $b \sin \theta_2$ and $b \sin \theta_3$, where $a \cos \theta_3$ and $b \sin \theta_2$ are imaginary.

Hence $a \cos \theta$, $b \sin \theta$, where θ can have all values, real or purely imaginary, are the coordinates of points on the real, the $(1, i)$ and the $(i, 1)$ branches of the curve.

(b) Construct the real branch, the (α, β) branch* [see Art. 127] and the (β, α) branch. Let a' and b' be the semi-conjugate diameters OA and OB of the curve corresponding to these branches so that the equation

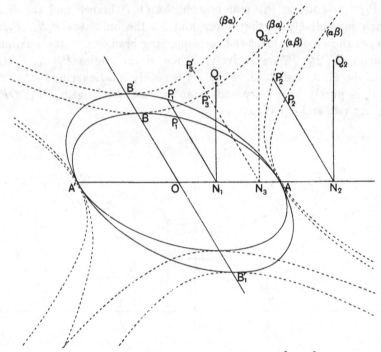

of the curve referred to these diameters as axes is $\dfrac{x^2}{a'^2} + \dfrac{y^2}{b'^2} = 1$. On OB take two points B' and B_1' at distances equal to OA from O. Describe an ellipse through A, B', A', B_1' to touch the given ellipse at A. The equation of this ellipse is $x^2 + y^2 = a'^2$. Construct the imaginary branches of this ellipse corresponding to the axes OA and OB.

Take P_1, P_2, P_3 points on the real branch, the (α, β) branch and the (β, α) branch respectively of the ellipse, and let the ordinates $P_1 N_1$, $P_2 N_2$, $P_3 N_3$ at these points meet the corresponding branches of the ellipse $x^2 + y^2 = a'^2$ in P_1', P_2', P_3' respectively. ON_1 and ON_2 are real

---

* This is the branch for which the coordinates are x, iy, the axes of coordinates being the conjugate diameters of the curve, which make angles α and β with the major axis.

and ON_3 is purely imaginary. Through N_1, N_2, and N_3 erect perpendiculars N_1Q_1, N_2Q_2, and N_3Q_3 equal respectively to N_1P_1', N_2P_2', and N_3P_3', the first and third of which are real and the second purely imaginary. Let Q_1ON_1, Q_2ON_2, and Q_3ON_3 be θ_1, θ_2, and θ_3 respectively.

Then $\qquad OQ_1^2 = ON_1^2 + N_1Q_1^2 = ON_1^2 + N_1P_1'^2 = a'^2.$

Similarly $\qquad\qquad OQ_2^2 = OQ_3^2 = a'^2.$

Therefore ON_1, ON_2, ON_3 are respectively $a'\cos\theta_1$, $a'\cos\theta_2$, and $a'\cos\theta_3$. Hence from the equation to the ellipse P_1N_1, P_2N_2, P_3N_3 are respectively $b\sin\theta_1$, $b\sin\theta_2$, and $b\sin\theta_3$.

Hence $a\cos\theta$, $b\sin\theta$, where θ can have all values real or purely imaginary, are the coordinates of points on the real, the (α, β) and the (β, α) branches of the curve, the axes of coordinates being inclined at an angle $\beta - \alpha$. If the major axis is taken as the initial line the eccentric angle of a point on the (α, β) branch is of the form $(\alpha + \theta_i)$. Hence, if α be constant, points on this branch are obtained by varying θ_i, and the different (α, β) figures are obtained by varying α.

The locus of Q_1, Q_2, Q_3, ... is a circle described on AA' as diameter.

(c) Similarly the coordinates of a point on the hyperbolae

$$\frac{x^2}{a^2} - \frac{y^2}{b^2} = 1 \quad\text{and}\quad -\frac{x^2}{a^2} + \frac{y^2}{b^2} = 1$$

may be expressed in the form $a\cos\theta$, $ib\sin\theta$ and $ia\cos\theta$, $b\sin\theta$, where θ can have any value, real or purely imaginary. If the axes are inclined at an angle ω the points lie on (α, β) and (β, α) branches, where α and β give the pair of conjugate diameters, which are inclined at an angle ω.

Consider generally what is represented by the point

$$a\cos(\theta + s\theta_1),\ b\sin(\theta + s\theta_1).$$

These coordinates may be written

$$a\cos\theta\cos s\theta_1 - a\sin\theta\sin s\theta_1,\ b\sin\theta\cos s\theta_1 + b\cos\theta\sin s\theta_1.$$

Now, if $\cos s\theta_1$ is real, $\sin s\theta_1$ is imaginary and vice versa. Considering the real and imaginary parts of these coordinates and assuming that $\cos s\theta_1$ is real, a point is given by a real length

$$\sqrt{a^2\cos^2\theta + b^2\sin^2\theta}\ .\ \cos s\theta_1,$$

measured in a direction making an angle $\tan^{-1}\dfrac{b\sin\theta}{a\cos\theta}$ with the axis of x, and by an imaginary length

$$\sqrt{a^2\sin^2\theta + b^2\cos^2\theta}\ .\ \sin s\theta_1,$$

measured in a direction making an angle $\tan^{-1} - \dfrac{b \cos \theta}{a \sin \theta}$ with the axis of x. $a^2 \cos^2 \theta + b^2 \sin^2 \theta$ and $a^2 \sin^2 \theta + b^2 \cos^2 \theta$ are the squares of semi-conjugate diameters of an ellipse of semi-axes a and b, the eccentric angle of the end of a diameter being θ. Their directions are also the direction of a pair of conjugate diameters, for if m and m' be the tangents of the angles which they make with the axis of x, then

$$mm' = - \frac{b^2}{a^2} \frac{\sin \theta \cos \theta}{\cos \theta \sin \theta} = - \frac{b^2}{a^2}.$$

Hence if $s\theta_1 \equiv 0$, the real branch of an ellipse of semi-axes a and b is given; if $\theta \equiv 0$, the $(1, i)$ and $(i, 1)$ branches are given; if θ has a constant value the (θ, ϕ) and (ϕ, θ) branches of the ellipse are given, where ϕ is the angle which the diameter conjugate to the diameter given by θ makes with the axis of x.

75. *Evaluation of integrals.*

If $\qquad \sin \theta = \dfrac{ax + h}{\sqrt{h^2 - ab}}$, then $\cos \theta = \dfrac{i \sqrt{a}}{\sqrt{h^2 - ab}} \sqrt{ax^2 + 2hx + b}$,

$$\tan \theta = \frac{1}{i \sqrt{a}} \frac{ax + h}{\sqrt{ax^2 + 2hx + b}} \quad \text{and} \quad \frac{d\theta}{dx} = \frac{-i \sqrt{a}}{\sqrt{ax^2 + 2hx + b}},$$

whatever are the values of a, h, b and x, provided a is not zero. These results may of course be applied for the evaluation of integrals. Thus if it is required to evaluate the integral

$$\int f (\sqrt{ax^2 + 2hx + b}) \, dx,$$

this integral may at once be written as

$$\int f \left(\frac{\sqrt{h^2 - ab} \cos \theta}{i \sqrt{a}} \right) \frac{\sqrt{h^2 - ab}}{a} \cos \theta \, d\theta.$$

An integral of the form

$$\int f (ax^2 + 2hx + b, \, x) \, dx$$

becomes at once

$$\int f \left(\frac{\sqrt{h^2 - ab}}{i \sqrt{a}} \cos \theta, \, \frac{\sqrt{h^2 - ab} \sin \theta - b}{a} \right) \cdot \frac{\sqrt{h^2 - ab}}{a} \cos \theta \, d\theta.$$

It may be noticed that

$$\frac{d \sin i\theta}{d\theta} = i \sin \left(i\theta + \frac{\pi}{2} \right),$$

$$\frac{d \cos i\theta}{d\theta} = i \cos \left(i\theta + \frac{\pi}{2} \right).$$

76. Tracing of imaginary straight lines.

Let the equation of the pair of lines considered be

$$y^2 + m^2 x^2 = 0, \quad \dots \dots \dots \dots \dots \dots (1)$$

the axes being rectangular.

(1) Draw a system of straight lines parallel to the axis of y. Any one of these straight lines PNP' will meet the axis of x in a real point N, and the pair of straight lines in a pair of conjugate imaginary points P and P'. The locus of the points P and P' so obtained is a pair of lines OP and OP', which make angles $\tan^{-1}(\pm\,im)$ with the axis of x.

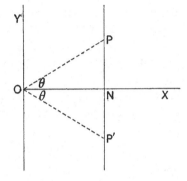

(2) Change the axes of coordinates so that $x \sin \alpha - y \cos \alpha = 0$ and $x \sin \beta - y \cos \beta = 0$ are the axes of x and y respectively, the angles α and β being real.

The equation of the pair of lines is then

$$X^2 (\sin^2 \alpha + m^2 \cos^2 \alpha) + Y^2 (\sin^2 \beta + m^2 \cos^2 \beta)$$
$$+ 2XY (\sin \alpha \sin \beta + m^2 \cos \alpha \cos \beta) = 0.$$

Let $\tan \alpha \tan \beta = - m^2$. Then the equation is

$$Y^2 + X^2 \frac{\sin^2 \alpha + m^2 \cos^2 \alpha}{\sin^2 \beta + m^2 \cos^2 \beta} = 0$$

or
$$Y^2 - X^2 \frac{\sin 2\alpha}{\sin 2\beta} = 0$$

or
$$Y^2 - X^2 \frac{M - m^2 M'}{M' - m^2 M}$$

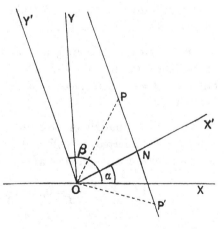

if $\tan \alpha = M$ and $\tan \beta = M'$. Hence if $m^2 = 1$ the equation becomes $Y^2 + X^2 = 0$.

The locus of the points of intersection of the given lines with real lines parallel to the new axis of y, can as in the previous case be shown to be a pair of lines OP and OP' which pass through O, P and P' being conjugate imaginary points.

If $\tan \alpha \tan \beta = - m^2$ the lines $y - x \tan \alpha = 0$ and $y - x \tan \beta = 0$ are harmonic conjugates of the lines $y^2 + m^2x^2 = 0$. They are also a pair of conjugate diameters of the ellipse $\dfrac{x^2}{a^2} + \dfrac{y^2}{b^2} = 1$ if $\dfrac{b}{a} = m$.

Hence the graphic representation of a pair of conjugate imaginary lines is as follows. The pair of conjugate imaginary lines are the double rays of a real overlapping involution pencil. All pairs of real conjugate rays of the pencil are harmonic conjugates of the pair of given lines. If any pair of these conjugate rays be taken as axes of coordinates the graph of the pair of imaginary lines with respect to these axes is a pair of straight lines through their real point. By varying the pair of conjugate rays which are taken as axes all points on the given straight lines can be graphically represented.

To construct the point of intersection of a real line with an imaginary line.

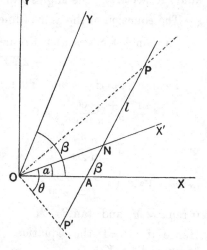

Take as axes of coordinates the bisectors of the angles between the imaginary line and its conjugate imaginary line. Let the combined equation of these lines be

$$y^2 + m^2x^2 = 0.$$

Let the given real line (l) make an angle β with the axis of x.

Draw through O a line, making an angle α with the axis of x, where $\tan \alpha \tan \beta = - m^2$, to meet l in N. Let $ON = h$.

Then the equation of the lines $y^2 + m^2x^2 = 0$ referred to ON and a line parallel to l through O as axes is

$$Y^2 - X^2 \frac{\sin 2\alpha}{\sin 2\beta} = 0, \quad \text{or} \quad \frac{Y}{X} = \pm i \sqrt{- \frac{\sin 2\alpha}{\sin 2\beta}}.$$

Therefore $\qquad NP = -NP' = -ih \sqrt{- \dfrac{\sin 2\alpha}{\sin 2\beta}}.$

Hence P and P' are a pair of conjugate imaginary points and for lines parallel to l they lie on the lines OP and OP'.

The required points P and P' are the points of intersection of l with the graph of the lines, for which the imaginary axis is parallel to l and the real axis is the harmonic conjugate of this line with respect to the pair of imaginary lines.

If the point A where the line l meets the axis of X is fixed and $OA = K$, it may be easily deduced that, when β varies, the locus of N is the ellipse

$$\frac{\left(x - \dfrac{K}{2}\right)^2}{\left(\dfrac{K}{2}\right)^2} + \frac{y^2}{\left(m\dfrac{K}{2}\right)^2} = 1.$$

If the values $x_1 + ix_2$, $x_1 \tan a + ix_2 \tan \beta$ [see Arts. 129 and 132] are substituted for x and y in $y^2 + m^2 x^2 = 0$, it follows at once that

$$\tan a \tan \beta = -m^2 \text{ ,and } \frac{x_1^2}{x_2^2} = -\frac{\tan \beta}{\tan a}$$

Therefore
$$\frac{x_1}{x_2} = \pm \frac{m}{\tan a} = \mp \frac{\tan \beta}{m}.$$

77. *To find the imaginary angle, which the line OP' in the preceding makes with the axis of x.*

Let $\beta - a = \omega$ and $P'OX'$ be θ.

Then from the equation of the line

$$\sqrt{\frac{\sin 2a}{\sin 2\beta}} = \frac{P'N}{ON} = \frac{\sin \theta}{\sin (\omega + \theta)}.$$

Therefore
$$\sin \omega \cot \theta + \cos \omega = \sqrt{\frac{\sin 2\beta}{\sin 2a}}.$$

Now θ will generally be a complex angle. Let $\theta = \theta_1 + s\theta_2$. Then

$$\tan \theta = \tan (\theta_1 + s\theta_2) = \frac{\tan \theta_1 + \tan s\theta_2}{1 - \tan \theta_1 \tan s\theta_2}.$$

Therefore
$$\sin \omega \frac{1 - \tan \theta_1 \tan s\theta_2}{\tan \theta_1 + \tan s\theta_2} + \cos \omega = \sqrt{\frac{\sin 2\beta}{\sin 2a}}.$$

Equating real and imaginary parts, since $\sqrt{\dfrac{\sin 2\beta}{\sin 2a}}$ is imaginary,

$$\sin \omega + \cos \omega \tan \theta_1 = \sqrt{\frac{\sin 2\beta}{\sin 2a}} \tan s\theta_2, \quad \dots\dots\dots\dots\dots(1)$$

and
$$(\cos \omega - \sin \omega \tan \theta_1) \tan s\theta_2 = \tan \theta_1 \sqrt{\frac{\sin 2\beta}{\sin 2a}}. \quad \dots\dots\dots\dots(2)$$

Therefore $\quad (\sin \omega + \cos \omega \tan \theta_1)(\cos \omega - \sin \omega \tan \theta_1) = \tan \theta_1 \dfrac{\sin 2\beta}{\sin 2a},$

or $\quad\quad \sin(\omega + \theta_1)\cos(\omega + \theta_1) = \sin \theta_1 \cos \theta_1 \dfrac{\sin 2\beta}{\sin 2a}.$

Therefore $\quad\quad \sin 2a \sin\{2\omega + 2\theta_1\} = \sin 2\theta_1 \sin 2\beta,$

whence it follows that $\theta_1 = a$.

Let $\theta_1 = a$ in (1), then

$$\sin \omega + \cos \omega \tan a = \sqrt{\dfrac{\sin 2\beta}{\sin 2a}} \tan s\theta_2.$$

Therefore $\quad\quad\quad\quad \sqrt{\dfrac{\sin a}{\cos a} \times \dfrac{\sin \beta}{\cos \beta}} = \tan s\theta_2.$

Therefore $\quad\quad\quad\quad im = \tan s\theta_2.$

Hence the angle $\quad P'OA = NOA - NOP'$

$$= a - (a + \tan^{-1} im)$$

$$= -\tan^{-1}(im) = \tan(-im).$$

This is the angle which the line $y + imx = 0$ makes with the axis of x.

It will be noticed that in the preceding the positive sign was given to $\sqrt{\dfrac{\sin 2\beta}{\sin 2a}}$ in the first equation. The reason for this is set forth in Art. 126.

78.   Critical lines and the circular points at infinity.

Consider the double points E and E' of an overlapping involution, which is orthogonal at S.   If SO be the perpendicular from S on the base, then

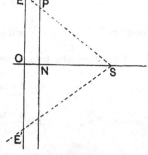

$$OE = -OE' = i.SO.$$

Hence if P be any point on SE and N the foot of the perpendicular from P on SO,

$$SO^2 + OE^2 = 0 \text{ and } SN^2 + PN^2 = 0.$$

Hence the measure of the distance of any point on one of these lines, SE and SE', from the point S in which they intersect (or from any other point on the lines) is zero. These lines are termed *the critical lines* of the point (see Art. 22). They are the locus of all points in the plane the measure of whose distances from S is zero.

Since the measure of the distance between any two points on one of these lines is zero, the definition of the sine and cosine (and hence also of the tangent) of the angle between two lines fails, when one of them is a critical line.

It may be noticed that even the formula $c^2 = a^2 + b^2 - 2ab \cos C$ fails for a real triangle in certain cases. Let A, B be two real points at a finite distance apart. Through A and B draw two parallel lines AC, BC, to make given equal angles with AB. Then AC and BC meet at infinity.

In the triangle ABC, by the given formula

$$AB^2 = AC^2 + BC^2 - 2AC \cdot BC \cos C$$
$$= (BC - AC)^2, \text{ since } \cos C = 1.$$

But, if $BC - AC$ represents anything, it represents the distance BN, where AN is the perpendicular from A on BC.

Hence $AB^2 = BN^2$.

This obviously is not true unless C lies on AB. Since even the formula quoted fails in certain cases for a real triangle it is not a matter of surprise that it and other formulae should fail for imaginary points, lines, and angles in certain cases.

Generally a real or an imaginary straight line meets the critical lines of different points in the plane in different pairs of points. But this is not the case with "the line at infinity." On the line at infinity, regarding the region at infinity as a straight line, all involutions of orthogonal lines through different vertices determine the same involution and therefore the critical lines of different points meet the "line at infinity" in the same pair of imaginary points. Hence the critical lines of a point may be regarded as the connectors of the point in question to this pair of points which have been termed the circular points at infinity, but may be better called the critical points of the plane. (See Art. 22.)

For the graphic representation of the critical lines of a point see Arts. 76 and 147.

Hence unless two points are situated on a critical line it follows that, *if the measure of the distance between them is zero, they must coincide.*

Analytical. The equation of the pair of bisectors of the angles between the lines $x^2 + y^2 = 0$ is $\dfrac{x^2 - y^2}{1 - 1} = \dfrac{xy}{0}$. Hence the bisectors of the angles between a pair of critical lines are indeterminate.

It may be noticed that, if it be assumed that the addition formula holds for the tangents of angles involving the angle $s\dfrac{\pi}{4}$, then whatever angle a may be

$$\tan\left(a + s\,\frac{\pi}{4}\right) = i, \text{ unless } \tan a + i = 0.$$

The value of y, which corresponds to the value $s\,\dfrac{\pi}{4}$ of $s\theta$ in the formula, $\tan s\theta = \dfrac{e^v - e^{-v}}{e^v + e^{-v}}$, is infinite. Hence a finite addition to y does not generally make any difference in the value of y.

The critical lines through a point have an equation of the same form whatever pair of lines at right angles through the point are taken as axes of coordinates (see Art. 76). Let the coordinates of any point referred to rectangular axes be x and y and those referred to another pair of axes x' and y', the two systems of axes being inclined at an angle a. The equations of the critical lines of the origin referred to the first axes are $x + iy = 0$ and $x - iy = 0$ and those referred to the second axes are $x' + iy' = 0$ and $x' - iy' = 0$. If the second axes are rotated through an angle a by means of the usual substitutions, the first equations are obtained from the second.

The critical points in a plane and the critical lines generally are considered in Art. 147.

79. *The perpendicular from a given point $x'\,y'$ on an imaginary line $ax + by + c = 0$.*

In the same way as when the line and point are real it may be proved that the perpendicular from the point $x'\,y'$ on the line is

$$\frac{ax' + by' + c}{\sqrt{a^2 + b^2}}$$

Equation of the bisectors of the angles between a pair of conjugate imaginary lines.

Let the equations of the lines be

$$ax + by + c + i\,(a'x + b'y + c') = 0$$

and

$$ax + by + c - i\,(a'x + b'y + c') = 0,$$

or shortly

$$L + iL' = 0 \quad \text{and} \quad L - iL' = 0.$$

Equating the perpendiculars from any point on the two lines and squaring, the equation is found to be

$$\frac{L^2 - L'^2}{a^2 + b^2 - (a'^2 + b'^2)} = \frac{LL'}{aa' + bb'},$$

or

$$\frac{(ax + by + c)^2 - (a'x + b'y + c')^2}{(a^2 + b^2) - (a'^2 + b'^2)} = \frac{(ax + by + c)(a'x + b'y + c')}{aa' + bb'} \quad . \quad \ldots\ldots(1)$$

If $c \equiv c' = 0$, the combined equation of the lines is

$$(ax + by)^2 + (a'x + b'y)^2 = 0$$

and by the usual formula the equation of the bisectors of the angles between these lines is

$$\frac{x^2 - y^2}{a^2 + a'^2 - b^2 - b'^2} = \frac{xy}{ab + a'b'}.$$

If c and c' are made equal to zero in (1) the two equations may be shown to be identical.

80. Systems of lines through a point.

From the earlier articles of this chapter it will be seen that a real point contains:

(1) an infinite number of real lines determined by real angles measured from a base line drawn through the point:

(2)   an infinite number of what may be termed purely imaginary lines determined by angles, which are entirely imaginary, measured from the base line:

(3)   an infinite number of infinite systems of imaginary lines—whose determining angles with reference to the base line are complex angles, i.e., angles which may be expressed as the sum or difference of real and purely imaginary angles. Each infinite system may be obtained by measuring imaginary angles from some real line of (1), or by measuring real angles from some line of (2).

The system of lines through an imaginary point is of the same nature except for the fact that the point itself is imaginary.

The above should be compared with the statement as to points on a real line given at the commencement of Art. 51.

Systems of lines through a real and through an imaginary point considered in reference to the critical lines.   Through every real point there passes:

(1)   a pair of lines termed the critical lines of the point. These lines are the connectors of the point to the circular points at infinity:

(2)   an infinite number of pairs of real lines, which are harmonic conjugates of the critical lines and are therefore at right angles, also an infinite number of pairs of imaginary lines, which are harmonic conjugates of the critical lines and therefore at right angles. Such pairs of imaginary lines are not generally pairs of conjugate imaginary lines.

Since every line through a point has a conjugate in every involution at the point it follows that (2) includes all the lines through the point.

(3)   an infinite number of pairs of conjugate imaginary lines. Each pair of such conjugate imaginary lines has a pair of the real orthogonal lines of (2) for the internal and external bisectors of the angles between them. Those pairs of conjugate imaginary lines, which have the same pair of real orthogonal lines for bisectors, form an involution of which the real bisectors are the double rays. The critical lines being harmonic conjugates of all pairs of real orthogonal lines through the point are a pair of conjugate rays of all such involution pencils.

Through every imaginary point there passes:

(1)   a pair of critical lines which are the connectors of the point to the circular points at infinity:

(2)   one real line which is its connector to its conjugate imaginary point:

(3)   a line perpendicular to this line, which is real except in so far as it passes through the imaginary point, i.e. this line would be real if the imaginary point were looked upon as the origin:

(4)   an infinite number of pairs of lines which are harmonic conjugates of the critical lines and are therefore at right angles. These pairs of lines divide themselves up into two groups (*a*) those which would be real if the point were taken as origin and (*b*) those which would still be imaginary:

(5)   an infinite number of pairs of lines obtained by joining the point to pairs of conjugate imaginary points. These pairs of lines have the pairs of lines (*a*) of (4) for bisectors of the angles between them, and those pairs which have the same pair

of bisectors form an involution. The critical lines are a pair of conjugate lines of all these involutions.

Hence, as might be expected, the system of lines through an imaginary point is the same as that through a real point, except for the fact that the point through which they all pass is an imaginary point. This also follows from the fact that the origin may be regarded as either a real or an imaginary point. The same was shown to be the case for real and imaginary straight lines.

If $a+ib$, $h+ik$ be an imaginary point the equation of the real line through it is $\dfrac{x-a}{b}=\dfrac{y-h}{k}$ and the perpendicular line $\dfrac{x-a-ib}{k}+\dfrac{y-h-ik}{b}=0$.

81. Theorems connected with projection.

If five points A, B, C, D, E, are situated on an imaginary straight line, other than a critical line, and (ABCD) = (ABCE), then E and D coincide, AC, AD, etc., being the measures of the distances between the points.

The imaginary line can be rotated round its real point into coincidence with some real line through that point. In this position A, B, C, D, E, will coincide with points A_0, B_0, C_0, D_0, E_0 and the anharmonic ratio of any four of the points A, B, C, D, E will be equal to the anharmonic ratio of the corresponding four of the points A_0, B_0, C_0, D_0, E_0.

Hence since $(ABCD) = (ABCE)$,
$$(A_0B_0C_0D_0) = (A_0B_0C_0E_0).$$

Therefore D_0E_0 must be zero.

Hence the measure of DE must be zero.

Therefore unless D, E are on a critical line they must coincide (Art. 78). This result may also be obtained by equating the anharmonic ratios of the measures of the distances between the points.

Two projective ranges on an imaginary straight line have a pair of self-corresponding points, real, coincident, or imaginary.

This follows in a similar way from Art. 21 (*b*).

82. (1) *A pencil with an imaginary vertex is intersected by all real transversals in ranges which are equi-anharmonic.*

This has been proved in Art. 21.

(2) *A pencil with an imaginary vertex is intersected by all imaginary transversals in ranges which are equi-anharmonic.*

This theorem may be proved as in Arts. [10] and [11] of the *Principles of Projective Geometry*. The proof there given depends on the fact that the sides of a triangle are proportional to the sines of the opposite angles, and this theorem has been shown to be true for an imaginary triangle, provided that no side is a critical line.

The theorem may also be deduced from the general case of Menelaus' theorem (Art. 65) as follows.

Let S be the vertex of the pencil and let two imaginary transversals meet the rays in A, B, C, D, and A', B', C', D' respectively and intersect at O. Through S draw any line to meet the transversals in U and V. Let the ratios of A, B, C, D, A', B', C', D' and of S with respect to the vertices of the triangle OUV be a, b, c, d, a', b', c', d' and s respectively.

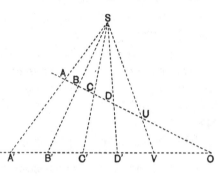

Then by Menelaus' theorem

$$aa's = 1, \quad bb's = 1, \quad cc's = 1, \text{ etc.}$$

Therefore

$$aa' = bb' = cc' = \ldots\ldots = \frac{1}{s}.$$

Therefore $(ABCD) = (A'B'C'D')$.

The point O is obviously a self-corresponding point of the ranges.

Correlatively *the pencils formed by projecting a range situated on an imaginary line from two imaginary points are equi-anharmonic.*

This may be proved in a similar way to the above by the general case of Ceva's theorem (Art. 65).

(3) *If two pencils with imaginary vertices have the connector of their vertices for a self-corresponding ray, all pairs of corresponding rays intersect on a fixed straight line.*

Let s be the line joining the vertices S and S' of the pencils. Let corresponding rays a, a' and b, b' intersect respectively at A and B. Join AB by the line u. Let d and d' a pair of corresponding rays meet u in D and D' and let SS' meet u in O.

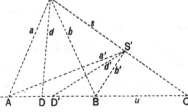

Then $(ABOD) = (ABOD')$.

Therefore D and D' coincide (Art. 81).

Conversely *if three pairs of corresponding rays intersect on a straight line, the connector of the vertices is a self-corresponding ray.*

In a similar manner it is possible to prove the correlative theorem, viz., *if the point of intersection of the bases of two projective ranges is a self-corresponding point, then the ranges are in plane perspective.*

Conversely *if the connectors of three pairs of corresponding points of two projective ranges are concurrent, the point of intersection of the bases is a self-corresponding point.*

If two projective pencils are such that for three pairs of corresponding rays the angles between the rays of one pencil are equal to the corresponding angles between the rays of the other pencil, the pencils are said to be equal and the angles between all pairs of corresponding rays are equal.

This follows from the preceding.

83. Real and imaginary correspondence.

If a pair of corresponding elements of two superposed projective forms are given by a relation
$$Axx' + Bx + Cx' + D = 0,$$
where x, x' determine pairs of corresponding elements, the correspondence is said to be real if A, B, C, D are real.

If pairs of corresponding elements are given by a relation
$$(A + iA')\, xx' + (B + iB')\, x + (C + iC'')\, x' + D + iD' = 0,$$
the correspondence is said to be imaginary.

(1) *Any pair of superposed projective pencils with a real vertex, and any pair of superposed projective ranges on a real base, have either two real elements of one which correspond to two real elements of the other or a pair of conjugate imaginary elements of one which correspond to a pair of conjugate imaginary elements of the other.*

Consider the correspondence of two superposed projective ranges on a real base.

This is given by
$$(A + iA')\, xx' + (B + iB')\, x + (C + iC')\, x' + (D + iD') = 0.$$

The real elements are given by
$$Axx' + Bx + Cx' + D = 0,$$
and
$$A'xx' + B'x + C'x' + D' = 0.$$

Therefore
$$(A'B - AB')\, x + (CA' - C'A)\, x' + (A'D - AD') = 0, \quad\ldots\ldots\ldots\ldots(1)$$
and
$$(BA' - B'A)\, x^2 + (BC' - B'C + DA' - D'A)\, x + (DC' - D'C) = 0.$$

The roots of this quadratic are either (1) a pair of real quantities or (2) a pair of conjugate imaginary quantities. From (1) these correspond to a pair of real quantities or to a pair of conjugate imaginary quantities.

The theorem for a pair of pencils is obtained by joining the points of the range to two real vertices.

(2) *Every pair of superposed projective pencils with a real vertex, and every pair of superposed projective ranges on a real base, have two elements of one which correspond to the conjugate imaginary elements of the other, either each to its own imaginary conjugate or each to the imaginary conjugate of the other.*

Write $y + iy'$ for x and $y - iy'$ for x'. Then equating real and imaginary parts it is found that

$$A\,(y^2 + y'^2) + (B + C)\,y + (C' - B')\,y' + D = 0,$$

and
$$A'\,(y^2 + y'^2) + (B' + C')\,y + (B - C)\,y' + D' = 0.$$

Hence
$$Ky + Ly' + M = 0, \quad\quad\quad\quad\quad\quad\quad\quad\quad\quad (1)$$

where

$$K = A'\,(B + C) - A\,(B' + C'), \quad L = A'\,(C' - B') + A\,(C - B), \quad M = A'D - AD',$$

and
$$y^2\,(K^2 + L^2) + y\,\{2KM + L\,(C^2 + C'^2 - B^2 - B'^2)\} + M^2$$
$$+ L\,\{D'\,(C' - B') + D\,(C - B)\} = 0. \quad\quad\quad (2)$$

(*a*) If the roots of (2) are real, viz., y_1 and y_2, the corresponding values of y' from (1) give a pair of conjugate imaginary points, and two pairs of conjugate imaginary points are corresponding points.

(*b*) If the roots of (2) are imaginary, viz., $(a + ib)$ and $(a - ib)$, let the corresponding values of y' from (1) be $(c + id)$ and $(c - id)$.

Then $(a + ib)$ and $(c + id)$ give as corresponding points

$$(a + ib) + i\,(c + id) \quad \text{and} \quad (a + ib) - i\,(c + id),$$

i.e.,
$$(a - d) + i\,(b + c) \quad \text{and} \quad (a + d) + i\,(b - c);$$

also $(a - ib)$ and $(c - id)$ give as corresponding points

$$(a - ib) + i\,(c - id) \quad \text{and} \quad (a - ib) - i\,(c - id),$$

or
$$(a + d) - i\,(b - c) \quad \text{and} \quad (a - d) - i\,(b + c).$$

Hence the theorem is proved.

84. The following are immediate consequences of Arts. 81 and 82 and may be proved in the same way that the corresponding theorems for real lines and points are proved in *The Principles of Projective Geometry*. They hold for all imaginary points and straight lines, excluding critical lines and pairs of points situated on the same critical lines.

   (i)   The properties of two triangles in perspective, Art. [13 (*a*)].

   (ii)   The harmonic property of the quadrangle and quadrilateral, Art. [45].

   (iii)   Construction of harmonic conjugates depending thereon, Art. [46].

   (iv)   The involution property of the quadrangle and quadrilateral, Art. [56].

Also the properties of involution ranges and pencils hold generally for pencils with imaginary vertices and for ranges on imaginary straight lines. Those for real involutions are set forth in Art. 7.

Formulae connecting the angles of an imaginary pencil.

Let a, b, c, d be any four concurrent lines real or imaginary (excluding critical lines) and let $\widehat{ab}$ denote the angle between the lines a and b. Then as in Art. [11] it may be shown that the anharmonic ratio of the pencil a, b, c, d is

$$\frac{\sin \widehat{ac}}{\sin \widehat{bc}} : \frac{\sin \widehat{ad}}{\sin \widehat{bd}} \equiv (abcd).$$

The relations connecting the angles, when the pencil is harmonic, may be proved as in *The Principles of Projective Geometry*.

85.   Projection of points into the circular points at infinity.

To project by a real projection any pair of conjugate imaginary points into the circular points at infinity.

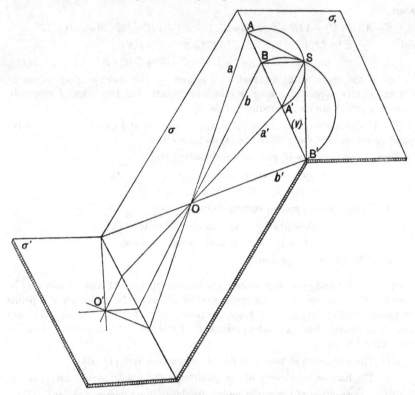

Let the given pair of conjugate imaginary points be the double points of the involution AA', BB', situated on the real line v in the plane σ. Through v draw any plane σ_1, and take any plane σ' parallel to σ_1 as plane of projection. On AA', BB', in the plane σ_1, describe two semi-circles to intersect at S. They will always intersect since the segments AA' and BB' overlap. Take S as centre of projection. Then the line v is projected into the line at infinity in the plane σ' and the involution AA', BB' is projected into the orthogonal involution on this line and its double points are the circular points at infinity.

To project a pair of real points into the circular points at infinity.

Let E and F, the two real points which are to be projected into the circular points at infinity, be situated on the real line v in the plane σ.

Through v draw any real plane σ_1, and let Ω and Ω' be the circular points at infinity in this plane.

Join $F\Omega$ and $E\Omega'$ to meet at S, and $F\Omega'$ and $E\Omega$ to meet at S'.

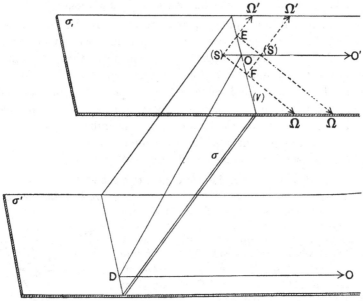

Then, if a plane σ' parallel to σ_1 be taken as plane of projection, and S or S' for centre of projection, E and F will be projected into Ω and Ω'. These points are the circular points at infinity in the plane σ' since the planes σ_1 and σ' are parallel.

In this projection v is projected into a real line, viz., the line at infinity. S and S' are a pair of conjugate imaginary points. Hence the line SS' is real and its point of intersection O with v is a real point and is projected into a real point O' at infinity. The connector to S of any other *real* point in the plane σ must meet the plane σ' in an imaginary point. For such connector, since it does not contain S', must be an imaginary line, and an imaginary line can only contain one real point. The real points on the line $\sigma\sigma'$ however remain real.

If D be any real point on the line $\sigma\sigma'$, the line DO corresponds to a real line in the plane σ', viz., DO'. Hence a pencil of real lines through O in the plane σ corresponds to a pencil of parallel real lines through O' in σ'. The real points on lines of one pencil correspond to imaginary points on the corresponding lines of the other pencil with the exception of O and O', which are both real, and the points on $\sigma\sigma'$ which are unaltered.

Hence, *if a pair of real points are projected into the circular points at infinity, all real points in the plane are projected into imaginary points with the exception of one point O, on the connector of the pair of real points, and points on the line of intersection of the planes.*

Consequently it is imaginary points on a locus in the plane σ which are projected into real points on the corresponding locus in the plane σ', and it is for such imaginary points in the plane σ' that theorems are proved, when they are deduced from the properties of real points in the plane σ. It is only if and when projective properties of real and of imaginary points have been shown to be identical, that the process of projecting real points into the circular points at infinity is justifiable as a means of obtaining a theorem for real points.

$SES'F$ is a semi-real square of the second kind.

86. Projection of a semi-real quadrangle into a semi-real square.

(a) Let the quadrangle be of the first kind. Let A, A', B, B' be the vertices and EFG the diagonal points triangle, which is real. Let AA' and BB' meet FG in K and L. Project the pair of conjugate imaginary points, which are the double points of the involution determined by FG and KL, into the circular points at infinity. Then in the new figure

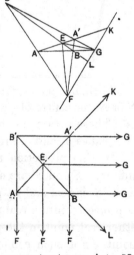

(1) the lines FAB', FBA' and FE are parallel, as are the lines GBA, $GA'B'$ and GE;

(2) the first three of these lines are perpendicular to the second three lines;

(3) the lines EK and EL are at right angles;

(4) the lengths BE and EB' are equal, as are the lengths AE and EA'.

Hence in the new figure the two pairs of conjugate imaginary points A, A' and B, B' are situated on two real lines EK and EL which are at right angles. Also $EB = B'E$ and $AE = EA'$. Also since the pencil $(E. KLFG)$ is harmonic and the lines EF and EG are at right angles, EF and EG are the bisectors of the angle AEB. But ABG is parallel to EG and therefore by symmetry (or see Arts. 53 and 66) $EA = EB$.

Therefore the points A, A', B, B' are the vertices of a semi-real square of the first kind.

(*b*) Let the quadrangle be of the second kind and let B and B' be the pair of real vertices. Then FG is the real side of the diagonal points triangle and E is its real vertex. F and G are a pair of conjugate imaginary points. Project these points into the circular points at infinity.

Then in the new figure

(1) the lines FAB', FBA' and FE are parallel, as are the lines GBA, $GA'B'$ and GE;

(2) the lines EK and EL are at right angles ;

(3) the lengths BE and EB' are equal, as are the lengths AE and EA'.

Hence in the new figure the pair of conjugate imaginary points A, A' and the pair of real points B and B' are situated on two real lines EK and EL which are at right angles. Also $EB = B'E$ and $AE = EA'$. Also the line ABG, which is parallel to a critical line EG, meets the real lines EA and EB through E, which are at right angles, in A and B. Therefore $EA = i \cdot EB$ [see Art. 78]. Hence A, A', B, B' is a semi-real square of the second kind.

Conversely it follows that *each of the four sides of a semi-real square of the second kind passes through a critical or circular point.*

Signs of trigonometrical functions.

In Art. 56 the sine, cosine and tangent of the angle between a real and an imaginary straight line were defined. In Art. 61 these definitions were extended to the case of the angle between a pair of imaginary straight lines. In these definitions, as far as the sine and cosine are concerned, there is an ambiguity of sign in respect to the denominator. This arises from the fact that there is a geometrical ambiguity—with real as well as with imaginary straight lines—in regard to the angle between two straight lines. There are four angles, less than 2π, between two given straight lines real or imaginary and by changing the sign of the square root in the expressions in Arts. 56 and 61 there are—both for real and for imaginary straight lines—four sets of values of $\sin \theta$, $\cos \theta$ and $\tan \theta$ corresponding to the four angles between the straight lines.

Certain conventions are laid down in the trigonometry of real lines as to which angles are those whose sine, cosine and tangent are represented by the expressions in Art. 56. It is convenient that the conventions introduced, when imaginary lines are considered, should be such as to make the sum and difference formulae, as used for real angles, apply when some or all of the angles considered are imaginary. The expression $\sqrt{OP^2 + OQ^2 - 2OP \cdot OQ \cos \omega}$ will in general be complex. It would seem that the correct convention is, in the case of the angle between two imaginary straight lines which has the smaller real part, to take the *same* value of

$$\sqrt{OP^2 + OQ^2 - 2OP \cdot OQ \cos \omega}$$

in the expressions for $\sin \theta$ and for $\cos \theta$. This has been done in Arts. 57 to 64, in which the sum and difference formulae for imaginary angles have been deduced. If different signs are given to this expression in the values of $\sin \theta$ and $\cos \theta$, the angle

$\pi - \theta$, positive or negative, is obtained. Generally it is believed that it will be found best to assume—as in the geometry of real lines—that the real part of $\sqrt{OP^2 + OQ^2 - 2OP \cdot OQ \cos \omega}$ is positive, when it exists, and to follow the conventions laid down in Arts. 71 and 72, when it is a purely imaginary quantity.

If the values of $\sin \theta_i$, $\cos \theta_i$, $\tan \theta_i$ are obtained by the method of Art. 72—in which the hypothenuse of the triangle is always real and positive—and if the positive axis of iy is taken in the negative direction of the real axis of y, it is found that the values in question are the same as those given on page 100 with the sign of i changed.

EXAMPLES

(1)   Prove that the sum of the measures of the distances of a point on an ellipse from the imaginary foci is equal to the minor axis.

(2)   Prove that the sum of the measures of the distances of a point on a hyperbola from the imaginary foci is equal to the transverse axis, the length of which is imaginary.

(3)   If two conjugate imaginary lines form a harmonic pencil with two real lines which are at right angles, then the real lines are the bisectors of the angles between the imaginary lines.

(4)   Prove that the ratio of the measures of the distances of a point on an ellipse from an imaginary focus and from the corresponding imaginary directrix is a constant quantity which is purely imaginary.

(5)   Prove that the equation of the chord joining two points on an ellipse determined by complex eccentric angles θ and ϕ is

$$\frac{x}{a} \cos \frac{\theta + \phi}{2} + \frac{y}{b} \sin \frac{\theta + \phi}{2} = \cos \frac{\theta - \phi}{2},$$

and that the equation of the tangent at the point is

$$\frac{x}{a} \cos \theta + \frac{y}{b} \sin \theta = 1.$$

(6)   Prove that the real and imaginary parts of the eccentric angle of the points where the line $y = mx + c$ meets the ellipse $\frac{x^2}{a^2} + \frac{y^2}{b^2} - 1 = 0$ are respectively $\tan^{-1} - \frac{b}{ma}$ and $\tan^{-1} \frac{-i \sqrt{c^2 - m^2 a^2 - b^2}}{c}$ if $c^2 > m^2 a^2 + b^2$.

(7)   Prove analytically that the real part of the angle between two imaginary straight lines through the origin is equal to one or other of the angles between the real bisectors of the angles between the lines and their conjugate imaginary lines.

(8)   Prove under the same circumstances that the imaginary part of the angle between the lines is the difference of the angles which the lines make with the bisectors in question.

CHAPTER IV

THE GENERAL CONIC

87 Definition of a conic.

Let S and S' be any two points real or imaginary; a, b, c three lines, real or imaginary, through S, and a', b', c' three lines, real or imaginary, through S'.

Then a, a', b, b', c, c' may be regarded as three pairs of corresponding rays of two projective or equi-anharmonic pencils with vertices at S and S', and the ray of one corresponding to any ray of the other may be constructed.

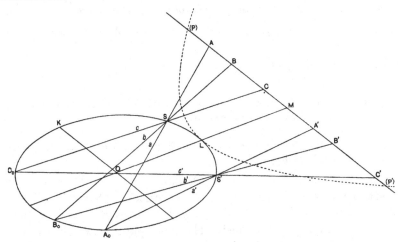

DEF. 5. *The locus of the points of intersection of corresponding rays of two projective pencils is a conic.*

It follows immediately from this definition that (1) a conic is completely determined when five points on it are given, and (2) that a conic may be described through any five given points.

The definition is illustrated by the figure. In this figure a real conic is determined by five real points S, S', A_0, B_0, C_0 and the pencils $(S . A_0 B_0 C_0)$ and $(S' . A_0 B_0 C_0)$ are constructed. P is any imaginary point on the conic. It can be constructed by means of the Poncelet figure corresponding to the diameter OK, which is parallel to the real line $AA'BB'CC'$ through the point P, and its conjugate diameter OL.

By definition the pencils $(S . A_0 B_0 C_0 P)$ and $(S' . A_0 B_0 C_0 P)$ are projective. This is the case if $(ABCP) = (A'B'C'P)$, where A, B, C and A', B', C' are the points in which the rays of the pencils meet the real line through P. That this is true may be verified by the figure. If P' be the conjugate imaginary point of P, then the point M, where OL meets the real line PP', is the mean point of P and P'.

If the angles a, β, γ, δ determine four points on a circle the anharmonic ratio of the pencil subtended by these points at the centre is $\dfrac{\sin (a - \gamma)}{\sin (\beta - \gamma)} : \dfrac{\sin (a - \delta)}{\sin (\beta - \delta)}$, and that of the pencil subtended at any point on the circle is $\dfrac{\sin \dfrac{a - \gamma}{2}}{\sin \dfrac{\beta - \gamma}{2}} : \dfrac{\sin \dfrac{a - \delta}{2}}{\sin \dfrac{\beta - \delta}{2}}$

Similarly, if a, β, γ, δ be the eccentric angles of points on the ellipse $\dfrac{x^2}{a^2} + \dfrac{y^2}{b^2} - 1 = 0$, the anharmonic ratio of the pencil subtended by these points at any point on the ellipse is $\dfrac{\sin \dfrac{a - \gamma}{2}}{\sin \dfrac{\beta - \gamma}{2}} : \dfrac{\sin \dfrac{a - \delta}{2}}{\sin \dfrac{\beta - \delta}{2}}$.

If a, β, γ, δ be complex angles, this expression still represents the anharmonic ratio of the four imaginary points on the ellipse (see Art. 74).

88. *Through five given points, no three of which are collinear, only one conic can be described.*

In Art. 87 the five points S, S', A, B, C completely determine the correspondence between the pencils whose vertices are S and S'. Hence the conic determined by the pencils with vertices S and S' is completely and uniquely determined. The only question which arises is whether the same conic is obtained if, of the five points, S, S', A, B, C, points other than S and S' are taken as the vertices of the pencils. That the same conic is obtained may be proved as follows.

To prove that whichever of the five given points, which determine a conic, are taken as the vertices of the two generating pencils, the same conic is obtained.

Let A, B, C, D, E be the five given points. Join A to C, D, E and B to C, D, E. Let a pair of corresponding rays of the pencils so determined meet at F. Join CF, and let CF meet BD, AD, BE and AE in K, L, M, N respectively. Then F is the second self-corresponding point of the ranges CKM and CLN and is a point on the conic determined by the pencils with vertices A and B.

Hence $(CFKM) = (CFLN)$.

Consider the pencils, DC, DB, DA and EC, EB, EA whose vertices are D and E. They determine ranges CKL and CMN on the line CF, and the second self-corresponding point of these ranges is the second point in which the conic determined by the pencils, vertices D and E, meets CF. Let F' be the second self-corresponding point.

Then $(CF'KL) = (CF'MN)$
which is the same as $(CF'KM) = (CF'LN)$.

But $(CFKM) = (CFLN)$.

Therefore F and F' coincide and the two conics meet the line CF in the same pair of points and are therefore the same conic.

Converse.

If A, B, C, D, E, F are six points on a conic the pencils subtended by the other four points at A and B are projective. But the five points A, C, D, E, F determine the conic and B may therefore be regarded as any point on the conic determined by these five points. Hence *at any point on a conic four given points on the conic determine a pencil of constant anharmonic ratio.*

Hence it follows that *a conic may be looked upon as the locus of a point at which four given points, which are on the conic, subtend a pencil of constant anharmonic ratio.* It is sometimes convenient to look upon a conic from this point of view.

89. *Every straight line, in the plane of a conic, meets the conic— unless it breaks up into a pair of straight lines—in two points which may be real or imaginary, and may be coincident.*

If any two points on a conic be joined to four other points on the conic, the pencils so formed determine two superposed projective ranges on every line in the plane. The conic meets each line in the self-corresponding points of these ranges, which are two points, real, imaginary or coincident (Art. 81).

There is one and only one tangent at every point on a conic, which does not break up into a pair of straight lines.

Let the conic be determined by five points A, B, C, P and O, real or imaginary. For every point P' on the curve $(P'.ABCP) = (O.ABCP)$.

Let P' be a point adjacent to P. Then $(P'.ABCP)=(O.ABCP)$.

Since $P'A$, $P'B$, $P'C$ and the anharmonic ratio of the pencil $(P'.ABCP)$ are given, the line PP' is a given line and is uniquely determined. This straight line is termed the tangent at P and contains all the points on the curve infinitely near to P.

If P' is real there will usually be a real point on the curve infinitely near to P and also an imaginary point infinitely near to P. Both of these points lie on the tangent at P.

90. *Given two tangents to a conic and their points of contact, to construct the tangent at any other point on the conic.*

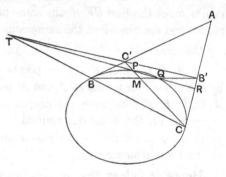

Let the tangents at B and C meet at A. Take P and Q any two points on the conic and let CP, BQ intersect at M and meet AB and AC in C' and B'.

Then
$$(B.BPQC)=(C.BPQC).$$
Therefore
$$(C'PMC)=(BMQB')=(B'QMB).$$

Hence $C'B'$, PQ and CB are concurrent at a point T. Let P and Q coincide. Then $C'B'$, BC and the tangent at P are concurrent. From the harmonic property of the quadrangle $TC'PB$ it follows that, if TP and AC meet at R, the range $ARB'C$ is harmonic.

Hence to construct the tangent at any point P, join BP to meet AC at B' and construct R the harmonic conjugate of A with respect to B', C. Then PR is the tangent at P.

Four fixed tangents to a conic determine ranges of constant anharmonic ratio on all other tangents to the conic.

Let AC and AB be any two tangents to the conic. Take points P_1, P_2, P_3, P_4 on the conic.

Let BP_1, BP_2, BP_3, BP_4 meet AC in B_1', B_2', B_3', B_4' and let the tangents at P_1, P_2, P_3, P_4 meet AC in the points R_1, R_2, R_3, R_4. Then the ranges $AR_1B_1'C$, $AR_2B_2'C$, $AR_3B_3'C$, $AR_4B_4'C$ are all harmonic. Hence $(R_1R_2R_3R_4)=(B_1'B_2'B_3'B_4')=$ the anharmonic ratio of the pencil formed by joining P_1, P_2, P_3, P_4 to any point on the conic = a constant for all positions of B on the conic.

91. *Correlative definition of a conic.*

The correlative definition of a conic is as follows :

The envelope of the lines joining pairs of corresponding points of two projective ranges is the correlative of a conic.

From this it follows by means of the correlative proofs

(1) *that the correlative of a conic can be described to touch five given lines :*

(2) *the anharmonic ratio of the range formed by the intersection of four fixed tangents to the correlative of a conic with a variable tangent is constant :*

(3) *only one correlative of a conic can be described to touch five given lines, no three of which are concurrent :*

(4) *through a given point two tangents real, coincident, or imaginary can be drawn to the correlative of a conic :*

(5) *every tangent to the correlative of a conic has one and only one point of contact.*

From Art. 90 and (2) above it follows that *the correlative of a conic is a conic.*

92. If, through the vertices of two projective pencils, a conic be described, these pencils determine on the conic ranges of points, which are termed *projective ranges on the conic.* Such ranges subtend projective pencils at every point on the conic.

The anharmonic ratio of four points A, B, C, D at any point on the conic is written $(ABCD)$. Hence the condition that A, B, C, D and A', B', C', D' should constitute two projective ranges on a conic is $(ABCD) = (A'B'C'D')$.

All pairs of superposed projective pencils have a pair of self-corresponding rays. Hence *every pair of projective ranges on a conic have a pair of self-corresponding points.* If the points of the two ranges be joined to any point on the conic, the self-corresponding rays of the pencils so formed meet the conic in the self-corresponding points of the projective ranges. Likewise, if through the vertex of an involution pencil a conic be described, the involution pencil determines on the conic a range of points which is termed *an involution range on the conic.*

The range in question is such that if the points which constitute it be joined to any point on the conic, the pencil so formed is an involution pencil.

It follows that, if three pairs of points A, A'; B, B'; C, C' are such that the anharmonic ratio of the pencil formed by joining four of them, not constituting two pairs, to a point on the conic, is equal to the anharmonic ratio of the pencil formed by joining the other points of the pairs to a point on the conic, then A, A'; B, B'; C, C' are pairs of conjugate points of an involution on the conic.

Since every involution pencil has a pair of double rays, every involution range on a conic has a pair of double points.

The correlative theorems and definitions, which hold for a real conic, are obviously true for the conic in general.

There can be no double point on a conic, which does not break up into two straight lines. For if possible let there be such a point P. Join P to any point Q on the conic. Then the line PQ meets the conic in three points, which is impossible unless the conic breaks up into a pair of straight lines.

93. It is now possible to prove for a conic in general the important theorems of projective geometry, which are proved for the real branch of a real conic in the *Principles of Projective Geometry*. The proofs are in most cases similar although in some cases they admit of simplification in the case of the general conic. They are shortly set forth in the following articles. The correlative theorem may in each case be proved by the correlative method.

Involution property of a conic.

A system of lines through any point S, real or imaginary, determines on any conic in their plane pairs of conjugate points of an involution.

Let S be the point and SAA', SBB', SCC' any three chords through it. Join AC' to meet SBB' in O.

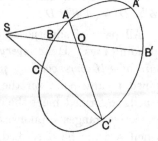

Then
$$(BB'CA) = (C'.BB'CA) = (BB'SO)$$
and
$$(B'BC'A') = (A.B'BC'A')'$$
$$= (B'BOS) = (BB'SO).$$

Therefore $(BB'CA) = (B'BC'A')$.

Hence (Art. 92) AA', BB', CC' are pairs of conjugate points of an involution on the conic.

If SAA', SBB' are regarded as fixed it is seen that SCC' is any chord through S.

Conversely *if AA', BB', CC' are three pairs of conjugate points of an involution on a conic, then AA', BB', CC' are concurrent.*

Similarly the correlative theorem and its converse may be proved.

94. Pole and polar.

If through a point S chords SAA', SBB', SCC', ... of a conic be drawn and L, M, N, ... are the harmonic con-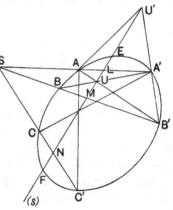
jugates of S with respect to AA', BB', CC', ..., then the locus of L, M, N is a straight line, which is termed the polar of S.

Join A to A', B', C', A, B, ..., and join A' to A, B, C, A', B', ... Then, since AA', BB', CC' form an involution, the pencils $(A . AA'BB'CC')$ and $(A' . A'A B'B C'C)$ are projective. But AA' is a self-corresponding ray of the pencils and therefore (Art. 49 (3)) the pencils are in plane perspective and pairs of corresponding rays intersect on a fixed straight line s.

Since s, in the figure, passes through U and U', etc., the points L, M, N in which s meets the chords through S are, by the harmonic property of the quadrangle, the harmonic conjugates of S with respect to A, A'; B, B'; C, C', ... Hence the locus of these harmonic conjugates is the straight line s.

The points where s meets the conic, viz. E and F, are the double points of the involution AA', BB', CC', ... They are the points of contact of the tangents from S to the conic and form a harmonic range on the conic with any pair of conjugate points A, A'; B, B'; etc.

The correlative theorem for the construction of the pole of a given line can be proved by the correlative method, which is given for the case of a circle in Art. [76].

Every real point in the plane of a real conic has a real polar with respect to the conic.

A real conic is a conic which is met by every real line in its plane in a pair of points, real, coincident, or conjugate imaginary. A conic with a real branch is always real as is also a conic with a real equation. The latter includes the former (Art. 106).

Through a real point S draw real chords SAA' and SBB' to meet the conic in real points (figure, Art. 94) or in pairs of conjugate imaginary points. These form the real or semi-real quadrangle $AA'B'B$, the diagonal points triangle of which has a real side UU'. The line UU' is the polar of S. By Art. 94 it is the locus of harmonic conjugates of S with respect to the points of intersection with the conic of all chords through S.

From the preceding it follows that the polar of a real point with respect to a conic, which has no real branch, such as the ellipse $\dfrac{x^2}{a^2} + \dfrac{y^2}{b^2} + 1 = 0$, is real.

95. *A conic determines on every straight line in its plane an involution, the double points of which are the points of intersection of the conic with the line.*

Let s be any straight line in the plane of the conic. On this straight line take points $P, Q, R, \ldots$. These points have each a polar with respect to the conic. Let these polars meet the line s in P', Q', R', ... respectively. Then P, P'; Q, Q'; R, R'; ... are harmonic conjugates of the points E and F in which the line s meets the conic. Therefore PP', QQ', RR', ... form an involution of which the points of intersection of s with the conic are the double points. If the conic and line are real the involution is real.

Correlatively, a conic determines at every point in its plane an involution pencil, the double rays of which are the tangents from the point to the conic. This involution pencil is real if the conic and point are real.

Two points P and P' are said to be conjugate points with respect to a conic if $(PP'QQ') = -1$, where PP' meets the conic in Q and Q'. The correlative definition holds for conjugate lines.

96. From Arts. 93—95 it follows that the properties of pole and polar hold not only for the real branch of a conic but for the conic in general. From the harmonic property of a point and its polar it follows that *every conic is in harmonic perspective with itself, any point and its polar being the centre and axis of perspective.* When the point is real, the polar is real, if the conic is real. In this case the real branch of the conic corresponds to itself and points on imaginary branches correspond to points on imaginary branches.

If the conic is real and an imaginary point be taken as centre of perspective and its imaginary polar as axis of perspective, real points on the curve correspond as a general rule to imaginary points.

97. Pascal's theorem.

The three points of intersection of the three pairs of opposite sides of a hexagon inscribed in a conic are collinear.

Let A, B, C and A', B', C' be any six points on a conic. If A, A'; B, B'; C, C' are looked upon as pairs of corresponding points, they determine two projective ranges on the conic.

Consider the pencils

$$(A' . ABC) \quad \text{and} \quad (A . A'B'C').$$

They have a self-corresponding ray AA' and are therefore in perspective.

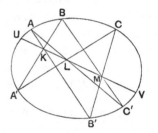

Hence the lines which join any pair of corresponding points of the ranges to A' and A intersect on a line s, which passes through K and L in the figure. This line meets the conic in the pair of self-corresponding points U and V of the projective ranges on the conic.

Similarly consider the pencils

$$(B' . ABC) \quad \text{and} \quad (B . A'B'C').$$

They have a self-corresponding ray BB' and are therefore in perspective. As in the previous case pairs of corresponding rays of the pencils intersect on a line s', which passes through K and M. This line s' meets the conic in the pair of self-corresponding points U and V of the projective ranges on the conic. Hence s' must coincide with s. Therefore K, L, M are collinear. But these are the points of intersection of the pairs of opposite sides of the hexagon $AB'CA'BC'$. Hence the theorem is proved.

98. Brianchon's theorem.

The correlative theorem, Brianchon's theorem, may be proved by the correlative method. The theorem is to the effect that, *the three connectors of the three pairs of opposite vertices of a hexagon circumscribed to a conic are concurrent.*

The converse of Pascal's theorem and also of Brianchon's theorem hold as for the real branch of a conic (see Art. [100]).

In the proof of Pascal's theorem, if A, A'; B, B'; C, C' be three pairs of conjugate imaginary points the Pascal line is real. For in the figure K, L, M are all points of intersection of pairs of conjugate imaginary lines and are therefore real.

If A, A'; B, B' are pairs of conjugate imaginary points and C and C' are real, then the Pascal line is imaginary. For K is real while L and M are imaginary. The real lines through L and M are the polars of AA' . CC' and BB' . CC'. Hence no other real lines can pass through L and M and therefore the Pascal line is imaginary, K being the real point on it.

Similarly if A, A are conjugate imaginary points and B, B', C, C' are real, the Pascal line is imaginary, M being the real point on it.

The condition that three imaginary points should lie on a real conic is that the hexagon formed by these points and their conjugate imaginary points (as opposite vertices) should be a Pascal hexagon. This condition is clearly satisfied if the real lines through the imaginary points are concurrent, for in this case the Pascal line is the polar of their point of intersection. Hence *three conjugate imaginary points, the real lines through which are concurrent, determine a real conic.*

99. Self-corresponding elements.

Determination of the self-corresponding elements of two superposed projective pencils or ranges.

Let a, b, c and a', b', c' be corresponding rays of two superposed pencils, vertex S. Describe a conic through S to meet the rays of the pencils in A, B, C, and A', B', C'.

Let the Pascal line of $AB'CA'BC'$ meet the conic in L and M. Then SL and SM are the self-corresponding rays of the two pencils.

The self-corresponding points of two projective ranges can be obtained by the correlative method or can be deduced from the preceding.

If two conics, real or imaginary, intersect in two points, they also intersect in two other points, real or imaginary (see page [272]).

Let S and S' be the given points of intersection of the conics (1) and (2). Take any three points A_0, B_0, C_0 on the conic (1). Join these points to S and S'. Then the conic (1) may be regarded as generated by the projective pencils ($S . A_0B_0C_0...$) and ($S' . A_0B_0C_0...$).

Let the rays of the first pencil meet the conic (2) in A, B, C, and let the rays of the second meet the conic (2) in A', B', C'. Then the self-

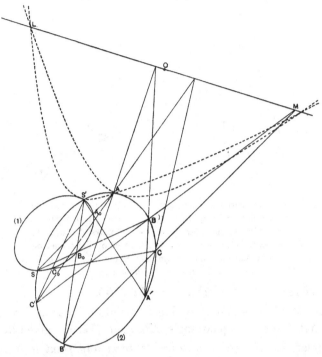

The figure represents the case in which the conics are real and intersect in two real points (given) and in a pair of conjugate imaginary points.

corresponding points of the ranges ABC and $A'B'C'$ are the points of intersection of the Pascal line of $AC'BA'CB'$ with the conic (2). Let these points be L and M. Then SL, $S'L$ and SM, $S'M$ are pairs of corresponding rays of the pencils, which generate the conic (1), and therefore the conic (1) passes through L and M, which are the two other points of intersection of the conics.

100. Desargues' theorem.

Every transversal is cut by a system of conics through four fixed points, in pairs of conjugate points of an involution.

Let any conic of the system of conics through the four fixed points Q, R, S, T meet any transversal s in P and P', and let opposite sides of the quadrangle meet s in A, A' and B, B' as in the figure.

Then $\quad\quad (PP'AB) = (Q.PP'AB) = (PP'TR),$

and $\quad\quad (P'PA'B') = (S.P'PA'B') = (P'PRT) = (PP'TR).$

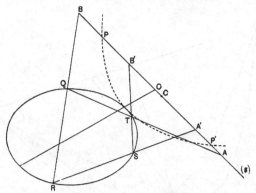

In the figure the transversal s, which is real, meets the
conic through Q, R, S, T, which are real, in a pair of
conjugate imaginary points. Real pairs of conjugate
points of the involution are situated on the same side
of C and imaginary conjugates on opposite sides of C,
where C is the centre of the involution.

Therefore $\quad\quad (PP'AB) = (P'PA'B').$

Therefore P, P' are a pair of conjugate points of the involution determined (Art. 84) by the quadrangle $QRST$ on s, i.e. of a given involution.

Correlatively *the pairs of tangents from any point to a system of
conics, touching four fixed lines, are pairs of conjugate rays of an involution pencil.*

This may be proved by the correlative method.

The converse theorems as stated in Art. [101] hold for the general
conic.

*One and only one conic, real or imaginary, can be described through
two given points A and B to determine pairs of conjugates of given
involutions on three given straight lines c, d, e. If the points A and B
are real, and likewise the involutions on c, d, e, then the conic is real.*

This can be proved by the method employed in Art. [114 (f)].

101. Carnot's theorem.

*If the points determined by a conic on the sides BC, CA, AB of any
triangle be A_1, A_2; B_1, B_2; C_1, C_2, then*

$$\frac{BA_1}{CA_1} \cdot \frac{BA_2}{CA_2} \cdot \frac{CB_1}{AB_1} \cdot \frac{CB_2}{AB_2} \cdot \frac{AC_1}{BC_1} \cdot \frac{AC_2}{BC_2} = 1.$$

Let the lines A_1B_1 and A_2B_2 meet the side BA in C_3 and C_4. Then C_1, C_2; A, B; and C_3, C_4 are pairs of conjugate points of an involution (Art. 84).

Therefore

$$\frac{AC_1}{BC_1} \cdot \frac{AC_2}{BC_2} = \frac{AC_3}{BC_3} \cdot \frac{AC_4}{BC_4}.$$

But by Menelaus' theorem

$$\frac{AC_3}{BC_3} = \frac{1}{\dfrac{CB_1}{AB_1} \dfrac{BA_1}{CA_1}},$$

and

$$\frac{AC_4}{BC_4} = \frac{1}{\dfrac{CB_2}{AB_2} \dfrac{BA_2}{CA_2}}.$$

Therefore $\dfrac{BA_1}{CA_1} \cdot \dfrac{BA_2}{CA_2} \cdot \dfrac{CB_1}{AB_1} \cdot \dfrac{CB_2}{AB_2} \cdot \dfrac{AC_1}{BC_1} \cdot \dfrac{AC_2}{BC_2} = 1.$

Correlatively *if the tangents from the vertices A, B, C of a triangle meet the opposite sides in three pairs of points A_1, A_2; B_1, B_2; C_1, C_2, then*
$$\frac{BA_1}{CA_1} \frac{BA_2}{CA_2} \frac{CB_1}{AB_1} \frac{CB_2}{AB_2} \frac{AC_1}{BC_1} \frac{AC_2}{BC_2} = 1.$$

This may be proved by the correlative method.

The converse theorems as stated in Art. [99] are all true.

102. *If two conics, real or imaginary, intersect in one point they also intersect in three other points.*

Let the conics S and S' intersect at A.

Let B be any other point on S.

Draw any chord through B meeting S' in Q and Q'.

Let AQ, AQ' meet S again in R and R'.

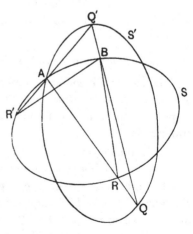

Then the pencils $B(Q)$, $B(R)$ have a $(1, 2)$ correspondence. For BQ determines BR and BR', and either BR or BR' uniquely determines BQ. Hence there are $2+1$ or 3 positions in which BQ and BR coincide (Art. [143]). For such positions of BR, Q and R coincide. But

Q and R can only coincide at a point of intersection of the conics S and S'. Hence in addition to A the conics S and S' intersect in three points.

103. *The locus of the common conjugates of points on a fixed line with respect to two conics is a conic.*

Let P be any point on the fixed line l and let L and L' be the poles of l with respect to the given conics. The polars of P with respect to the conics are two straight lines LQ and $L'Q$, which pass through L and L' and meet at a point Q, which is the common conjugate of P with respect to the conics. As P moves along l, LQ and $L'Q$ describe two pencils through L and L' which are each equi-anharmonic with the range described by P and are therefore projective with each other.

Hence the locus of Q is a conic, through L and L', and this conic is the locus of common conjugates of points on l with respect to the two conics.

The envelope of the common conjugates of lines through a fixed point with respect to two conics is a conic.

Let p be any line through the fixed point L, and let l and l' be the polars of L with respect to the given conics. The poles of p with respect to the conics are two points lq and $l'q$ which lie on l and l' and have for their connector q a line which is the common conjugate of p with respect to the conics. As p rotates round L, lq and $l'q$ describe two ranges on l and l' which are each equi-anharmonic with the pencil described by p and are therefore projective with each other.

Hence the envelope of q is a conic which touches l and l', and this conic is the envelope of common conjugates of lines through L with respect to the two conics.

104.   *Every pair of conics have a common self-conjugate triangle.*

The loci of the common conjugates of points on any two lines a and b with respect to the conics (1) and (2), are two conics.

The common conjugate of the point ab must be on both of these conics, which therefore intersect in one point. Therefore, by Art. 102, they intersect in three other points. Let P be one of these points. Let P_a and P_b be the

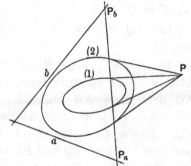

common conjugates of P on the lines a and b. Then P_aP_b is a common polar of P with respect to the two conics. Let P_aP_b meet the conics in A, A' and B, B'. The pair of common harmonic conjugates of A, A' and B, B' together with P form a common self-conjugate triangle of the conics.

It follows from the preceding that the locus of common conjugates of points on any straight line with respect to two conics passes through the vertices of their common self-conjugate triangle.

105. *Every pair of conics intersect in four points.*

Consider the locus of the common conjugates, with respect to the conics, of points on a line a which passes through one of the vertices F of their common self-conjugate triangle. The poles of the line a with

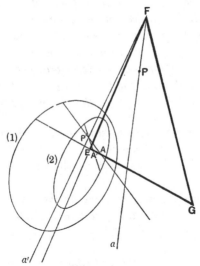

regard to the conics are two points A and A' which lie on EG, the side of the common self-conjugate triangle opposite to F. The polars of P any point on a pass through A and A' and meet at a point P'.

Since A, A' and also the points E, G, F are on the conic which is the locus of P', this conic must break up into a pair of lines, one of which is $EA'AG$ and the other FP'.

The polars of P' with respect to the conics pass through P. Therefore, if FP' be a', the locus of P for different positions of P' on a' is the line a. Hence the lines a and a' form an involution pencil. If FG is taken as a, then EF is a', so that FG and FE are conjugate elements of the involution. Consider the two double rays of this involution. For these P and P' lie on the same straight line. Therefore on these lines the two conics determine the same involution. Hence these lines meet the conics in the same pair of points.

Similarly there are a pair of lines through E and a pair of lines through G, which meet the conics in the same points. These six lines intersect in four points which are common to the two conics. Hence the conics intersect in four points.

By the correlative method it may be proved that *every pair of conics have four common tangents.*

Since the conics considered in Art. 103 intersect in four points it follows by Desargues' theorem that *the locus of common conjugates of points on a straight line is* (1) *the locus of common conjugates of all conics through the four points of intersection of this conic,* (2) *that the locus passes through the vertices of the common self-conjugate triangle,* (3) *that it is also the locus of the poles of the line with respect to conics of the system,* (4) *that it passes through the harmonic conjugates of the points, where the sides of the common inscribed quadrangle of the conics meet the given line, with respect to the vertices, and also through the double points of the involution determined by the quadrangle on the given line.*

No pair of conics real or imaginary can have more than one common self-conjugate triangle if they intersect in four distinct points.

This is proved in the same way as the corresponding theorem in Art. [124].

Hence *the diagonal points triangle of the common inscribed quadrangle of two conics and the diagonal triangle of their common circumscribed quadrilateral coincide.*

EXAMPLES

(1)   Prove that the imaginary tangents from an internal point to an ellipse are equally inclined to the focal distances of the point.

(2)   Prove that a pair of imaginary tangents from a real point to a conic subtend equal angles at a focus.

(3)   Prove that an imaginary tangent to a conic intercepts on two fixed real tangents a length which subtends a real constant angle at a real focus.

(4)   In Art. 94 if AA' and BB' are pairs of conjugate imaginary points and UU' meets SAA' and SBB' in A_0, B_0, prove that U and U' divide A_0B_0 in the ratio $\dfrac{1}{\sqrt{KK'}}$ (internally and externally), where $K = \dfrac{SM}{A_0M}$ and $K' = \dfrac{B_0M'}{SM'}$, M and M' being the points in which OAA' and OBB' are met by diameters conjugate to diameters parallel to these lines.

CHAPTER V

106.   Distinction between real and imaginary conics.

Hitherto the conic considered has been the general conic. *A real conic may be defined as a curve which is met by every real straight line in its plane in a pair of points, real, coincident or conjugate imaginary.*

In view of the correspondence between real points and purely imaginary points this definition may also be stated as follows. *A real conic is a curve which is met by every purely imaginary straight line in its plane in a pair of points, purely imaginary, coincident or conjugate imaginary.*

It follows that, if a real conic has a real point on it, it has a real branch.   For, if through the real point a real straight line be drawn, it must meet the conic in a second real point.

If a conic be described from data depending on imaginary points and quantities, another conic can be described in a similar way depending on the conjugate imaginary points and quantities.   The second conic is termed the conjugate imaginary conic of the first.   If a conic and its conjugate imaginary conic coincide, the conic in question must be a real conic.

When only the real rays of the generating pencils of the conic are taken into account, it is proved in the *Principles of Projective Geometry* that a conic determines on every real line in its plane a real involution. When real, the double points of this involution are the points of intersection of the line with the conic.   From Art. 95 it is seen that, when these double points are imaginary, they are also the points of intersection of the line and conic.   Being the double points of a real involution they are a pair of conjugate imaginary points.   Hence all conics considered in the *Principles of Projective Geometry* comply with the definition of a real conic.   If a conic of this nature can be described through five points or to satisfy other conditions, which determine the conic, the conic satisfying these conditions must be a real conic.

If a real conic passes through an imaginary point, it also passes through the conjugate imaginary point.

For let A be a point on the conic. Draw the real line l through A. This real line will meet the conic in the conjugate imaginary point of A, which is consequently on the conic.

A conic must be real if it passes through

    (*a*)   5 *real points,*

    (*b*)   3 *real points and a pair of conjugate imaginary points,*

    (*c*)   1 *real point and 2 pairs of conjugate imaginary points,* or

    (*d*)   3 *pairs of conjugate imaginary points.*

In Art. [149] real conics were described to comply with conditions (*a*), (*b*) and (*c*). Therefore a conic determined by these conditions is real. The fact that the conic is real in case (*d*) may be proved as follows. Let A, A'; B, B'; C, C', be three pairs of conjugate imaginary points on the conic. Then

$$(C . AA'BB') = (C' . AA'BB') = (C' . A'AB'B).$$

But if $(C . AA'BB') = K + iK'$, then $(C' . A'AB'B) = K - iK'$.

$$(K + iK') = (K - iK'), \therefore K' = 0 \text{ and } (C . AA'BB') \text{ is real.}$$

Hence the conic is real. (See Art. 23.)

This result may also be obtained as follows. The four points A, A', B, B', together with the point C, determine a conic.

The four points A', A, B', B, together with the point C', determine a second conic. This second conic is the conjugate imaginary conic of the first. But the six points A, A', B, B', C, C' all lie on a conic. Hence the conic and its conjugate imaginary conic coincide, and therefore the conic is real.

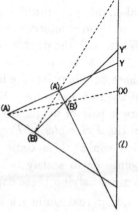

The conics (*a*) *which can be described through two pairs of conjugate imaginary points to touch a given real line and* (*b*) *correlatively the conics which can be described to touch two pairs of conjugate imaginary lines and to pass through a real point, are real.*

(*a*) Let AA' and BB' be the real sides of a quadrangle $AA'B'B$, and let AB', $A'B$ be a pair of conjugate imaginary sides of the quadrangle. Let the given real line l meet these pairs of the sides in X, X' and Y, Y'. A conic through A, B, A', B' which touches l will touch l at one of the double points of the involution determined by X, X', Y, Y'. These double points are the common harmonic conjugates of X, X' and Y, Y'. Since one pair of these are conjugate imaginary points their common harmonic conjugates—which are the double points of the involution—are real (Art. 8). Hence each of the two conics through A, A', B, B', which touches l, passes through a real point and therefore is real.

(b)   This is proved by the correlative method.

Of the conics, which pass through one real point, a pair of conjugate imaginary points, and touch two real straight lines, one pair is real and one pair is a pair conjugate imaginary conics.

Let K be the real point and A, A' a pair of conjugate imaginary points through which the conics are described, and l and m the two real straight lines which they touch.

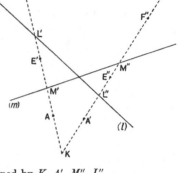

Then the lines KA and KA' (α' and α'') are a pair of conjugate imaginary lines. Hence the points in which they meet the real lines l and m are pairs of conjugate imaginary points, i.e. in the figure L', L'' and M', M'' are pairs of conjugate imaginary points. Hence the points A, L', M', are the conjugate imaginary points of A', L'', M'', respectively.

Hence the double points F'' and E' of the involution determined by K, A, M', L' are the conjugate imaginary points of the double points F''', E'' of the involution determined by K, A', M'', L''.

Here the lines $F''F''$ and $E'E''$ are real and they meet l and m in real points. Hence the two conics corresponding to these chords of contact are real. Art. [106 (b)].

The lines $F''E''$ and $E'F'''$ are conjugate imaginary lines and they meet l and m in two pairs of imaginary points, the first pair being conjugate imaginary points of the latter pair. Hence the two conics corresponding to these chords of contact are a pair of conjugate imaginary conics.

Of the conics which pass through a real point, a pair of conjugate imaginary points, and touch a pair of conjugate imaginary lines, one pair is real and one pair is a pair of conjugate imaginary conics.

In the proof of the last theorem let l and m be a pair of conjugate imaginary lines. Then M' and L'' are a pair of conjugate imaginary points as are also L' and M''. Hence the points A, M', L', are the conjugate imaginary points of A', L'', M''. Therefore the result follows as before.

The correlative theorems hold.

Since all circles pass through the circular points at infinity, a circle which passes through one real point and a pair of conjugate imaginary points, or through two pairs of conjugate imaginary points, is real.

Analytical.

Analytically a real conic is a conic whose equation does not explicitly contain "i".
The points of intersection of such a conic with a real straight line are given by a
quadratic equation, which does not explicitly involve "i", and whose roots are therefore real, coincident or conjugate imaginary. This definition includes as real conics
such curves as those given by the equations $x^2+y^2+a^2=0$ and $\frac{x^2}{a^2}+\frac{y^2}{b^2}+1=0$, although
there are no real points on the curves in question.

**107. Every imaginary conic contains one real or semi-real
quadrangle.**

Let S be the given conic and S' its conjugate imaginary conic. Let
any real line a meet S in P and P', and S' in Q and Q'. Then Q and
Q' are the conjugate imaginary points of P and P', and P, P' and Q,
Q' determine a real involution on a (Art. 8). Similarly the conics
determine real involutions on any other three real lines b, c and d.

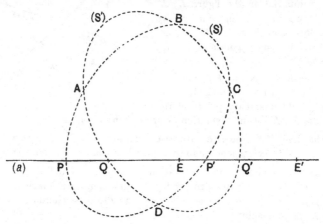

Let E and E' be a pair of real conjugates of the involution on a.
Describe a conic Σ through E and E' to determine conjugates of the
given involutions on b, c and d. This conic is real (Art. 99).

Let F and F' be a second pair of real conjugates of the involution
on a. Describe a second conic Σ' through F and F' to determine
conjugates of the given involutions on b, c and d. This conic also is
real.

The conics Σ and Σ' being real intersect in the vertices of a real or
semi-real quadrangle $ABCD$, and determine pairs of conjugates of the
involutions on a, b, c, d.

Describe a conic through P and A, B, C, D. It will pass through P'. Hence through P and P' there are two conics, viz. S and this new conic, which both pass through a pair of conjugates of the involutions on b, c and d. Hence these conics coincide and S passes through A, B, C, D. Therefore S contains a real or semi-real quadrangle.

The following is a particular case of the preceding.

If a conic be generated by pencils with real or with conjugate imaginary points for vertices, it contains another pair of real or conjugate imaginary points.

(a) Let the vertices of the projective pencils be real points S_1 and S_2. The pencils determine on any real straight line s two projective ranges in which a pair of real, or conjugate imaginary, points, A, A', correspond to a pair of real, or conjugate imaginary, points A_1 and A_1' (Art. 83 (1)).

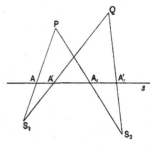

Let A and A' be real. Then in the figure P and Q are real.

Let A and A' be a pair of conjugate imaginary points. Then $S_1 A$ and $S_1 A'$ are conjugate imaginary lines as are $S_2 A_1$ and $S_2 A_1'$. Hence P and Q are a pair of conjugate imaginary points.

(b) Let the vertices S_1 and S_2 be a pair of conjugate imaginary points. Let A and A' be a pair of points which correspond to their conjugate imaginary points A_1 and A_1' (Art. 83 (2)). Then the pair of conjugate imaginary lines $S_1 A$ and $S_2 A_1$ meet in a real point P, and the pair of conjugate imaginary lines $S_1 A'$ and $S_2 A_1'$ meet in a real point Q. Hence P and Q are real.

Let A and A' be a pair of points each of which corresponds to the conjugate imaginary of the other, so that A and A_1', and A' and A_1 are pairs of conjugate imaginary points (Art. 83 (2)). Then $S_1 A$ and $S_2 A_1'$ are conjugate imaginary lines as are also $S_1 A'$ and $S_2 A_1$. Hence P and Q are a pair of conjugate imaginary points.

Analytical.

The most general form of the equation of a conic is

$$S + iS' = 0,$$

where $S = 0$ and $S' = 0$ are the general forms of the equation of a real conic.

Every imaginary conic whose equation is of the form $S + iS' = 0$ has a conjugate imaginary conic whose equation is $S - iS' = 0$ associated with it. If the first conic pass through an imaginary point, the second conic passes through the conjugate imaginary point. The points of intersection of these conics, through which the real conics $S = 0$ and $S' = 0$ pass, consist in general of either

(a) Four real points.

(b) Two pairs of conjugate imaginary points.

(c) Two real points and a pair of conjugate imaginary points.

Hence every imaginary conic contains four points, which constitute either a real quadrangle or a semi-real quadrangle of the first or second kind.

Every imaginary conic has a real or semi-real circumscribed quadri-lateral.

This may be proved by the correlative method to that employed to prove the first theorem of this article.

108. *No imaginary conic can have more than one self-conjugate triangle real or semi-real.*

Let ABC and $A'B'C'$ be two self-conjugate triangles of the conic.

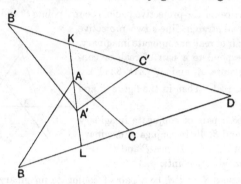

(*a*)   Let ABC and $A'B'C'$ be real.   Join AA' to meet BC and $B'C'$ in L and K.   Then A', K and A, L are pairs of conjugate points with respect to the conic, and the conic therefore meets AA' in a pair of points which are real or conjugate imaginary.   Similarly it meets BB' and CC' in pairs of points which are real or conjugate imaginary. Hence the conic is real.

(*b*)   Let ABC and $A'B'C'$ be semi-real.   Let A and A' be the real vertices.   Then as before the conic meets AA' in a pair of real or con-jugate imaginary points.   Produce BC and $B'C'$ to meet at D.   Then K, D and B', C' are pairs of conjugate points with respect to the conic; also B', C' is a pair of conjugate imaginary points.   Hence the conic meets $B'C'$ in a pair of real points (Art. 8).   Similarly it meets BC in a pair of real points.   Hence the conic is real.

(*c*)   If ABC is real and $A'B'C'$ is semi-real, it may be proved as in case (*b*) that the conic must be real.

109.   Construction of the self-conjugate triangle—real or semi-real—of an imaginary conic.

Let $ABCD$ be the real or semi-real inscribed quadrangle of the conic and let $abcd$ be the real or semi-real circumscribed quadrilateral.   Then the diagonal points triangle of *ABCD*, and the diagonal triangle of *abcd*, are both real or semi-real self-conjugate triangles of the conic. Hence they coincide in a real or semi-real triangle *EFG*. Therefore the sides of the quadrilateral *abcd* intersect in pairs on the sides of the triangle *EFG*. Also by the properties of pole and polar the points of contact of *a*, *b*, *c*, *d* are collinear in pairs with the vertices *E*, *F*, *G* of the triangle *EFG*.

Let ABCD be real or semi-real of the first kind.

Then the conic is in real harmonic perspective with any vertex of *EFG* for centre and the opposite side for axis of harmonic perspective.

Hence *A*, *B*, *C*, *D* are all real or are conjugate imaginary in pairs,
and *a*, *b*, *c*, *d*   „   „   „   „   „

If *A*, *B*, *C*, *D* are real and *a*, *b*, *c*, *d* are imaginary the conic is real. This is also the case if *A*, *B*, *C*, *D* are imaginary and *a*, *b*, *c*, *d* are real (Art. 106).

Therefore, *if the inscribed quadrangle is real or semi-real of the first kind, the circumscribed quadrilateral is real or semi-real of the first kind.*

Let ABCD be semi-real of the second kind.

Then the triangle *EFG* is semi-real. Hence the quadrilateral *abcd* has a semi-real diagonal triangle, and therefore it must be semi-real of the second kind.

Therefore *the inscribed quadrangle and the circumscribed quadrilateral of an imaginary conic are always of the same kind, real, semi-real of the first kind, or semi-real of the second kind.*

The conic cannot pass through any real points other than *A*, *B*, *C*, *D*, or through any pairs of conjugate imaginary points other than *A*, *B*, *C*, *D*, for if it did so it would be a real conic.

Likewise the conic cannot touch any real lines other than *a*, *b*, *c*, *d*, or any pairs of conjugate imaginary lines other than *a*, *b*, *c*, *d*, for if it did so it would be a real conic.

Hence the tangents at *A*, *B*, *C*, *D* and the points of contact of *a*, *b*, *c*, *d* are all imaginary. The tangents at *A*, *B*, *C*, *D* intersect in pairs of imaginary points in the sides of the triangle *EFG* and the points of contact of *a*, *b*, *c*, *d* are collinear in pairs with *E*, *F*, and *G*.

It should be noticed that *the real or semi-real quadrangle and quadrilateral of an imaginary conic are the common inscribed quadrangle and the common circumscribed quadrilateral of the conic and its conjugate imaginary conic.*

110. Geometrical origin of an imaginary conic.

Any imaginary conic can be described as the conic which circumscribes a real or semi-real quadrangle and touches either a real line or an imaginary line. In the latter case the quadrangle is semi-real of the second kind and the real point on the imaginary line touched by the conic lies on a side of the diagonal points triangle of the quadrangle.

Correlatively:

Any imaginary conic can be described as the conic which is inscribed in a real or semi-real quadrilateral and passes through either a real point, or an imaginary point. In the latter case the quadrilateral is semi-real of the second kind and the real line through the imaginary point, through which the conic is described, passes through a vertex of the diagonal triangle of the quadrilateral.

These results follow from the last article.

Through the vertices of a real or semi-real quadrangle two conics can be described to touch a real line. If these conics are imaginary they touch the real line at a pair of conjugate imaginary points and are therefore conjugate imaginary conics.

This result follows from Desargues' theorem.

Conics circumscribing a real or semi-real quadrangle so as to touch a given imaginary line.

Two conics can be described through the vertices of a real or semi-real quadrangle $ABCD$ to touch an imaginary line l. The points of contact are the double points of the involution determined by the sides of the quadrangle $ABCD$ on l. Let these points be E and F. Take l' the conjugate imaginary line of l. A conic through A, B, C, D may be described to touch this line at either of two points E' and F'. The points E, F, E', F' are two pairs of conjugate imaginary points. Hence the conic through A, B, C, D and E will be the conjugate imaginary conic of one of the conics through A, B, C, D and E' or F'. Similarly the conic through A, B, C, D and F will be the conjugate of the other one of these conics.

Similar results hold for the correlative construction of an imaginary conic.

111. Real conjugate points with respect to an imaginary conic.

Every real point in the plane of an imaginary conic has one real conjugate point with respect to the conic.

Let A be any real point. Describe a system of real conics through the vertices of the real or semi-real quadrangle contained by the given imaginary conic. Let the polars of A with respect to two of these real conics meet at A'. Then A and A' are real and are the double points of the involution determined on AA' by the conics described through the vertices of the quadrangle. Hence by Desargues' theorem they are conjugate points with respect to all conics—including the given imaginary conic—which pass through the vertices of the quadrangle.

The point A' is the one real point on the imaginary polar of A. It can be constructed as the point of intersection of the polars of A with respect to the given imaginary conic and its conjugate imaginary conic.

Correlatively:

Every real line in the plane of an imaginary conic has one real conjugate line with respect to this conic.

It follows from the above that the points of intersection of the real line AA' with the imaginary conic are a pair of conjugate points of a real involution. This involution is that determined on the line by the real or semi-real quadrangle of the conic.

The locus of the real conjugates of real points on a given real straight line with respect to an imaginary conic, is a conic (a) which passes through the vertices of the diagonal points triangle of the real or semi-real inscribed quadrangle of the conic, (b) which meets the sides of the quadrangle in six points which are the harmonic conjugates of the points, where the sides meet the given line, with respect to the vertices, and (c) which meets the given line in the double points of the involution determined on the line by the quadrangle. (Eleven points locus.)

Describe two real conics through the vertices of the real or semi-real inscribed quadrangle of the imaginary conic. The real conjugates with respect to the imaginary conic of real points on the given line are the common conjugates of these same points with respect to the two real conics. Hence the result follows from Art. [117 (2)].

In the proof in question two imaginary conics could have been substituted for the two real conics. Hence it follows that *the locus—which is the eleven points locus—is not only the locus—as proved in Art. [117 (2)]—of the real poles of the given line with respect to real conics of the system, but also that the imaginary points on the locus are the poles of the real line with respect to imaginary conics of the system.*

Cremona transformation.

The fact that every real point in the plane of an imaginary conic has one and only one real conjugate point with respect to the conic affords a method of deducing geometrical theorems from each other. The points are said to be derived from each other by what is termed the Cremona transformation.

If A' be the real conjugate of A, the relationship between A and A' is reciprocal. If A moves along a straight line a, the conjugate point A' describes a conic a', which passes through three fixed points E, F, G, which are the vertices of the real or semi-real self-conjugate triangle of the imaginary conic. Conversely if A describes a conic through E, F, G the conjugate point A' describes a straight line. If a corresponds to a' and b to b', the point ab corresponds to the point $a'b'$. To a straight line through any one of the points E, F, G corresponds a straight line through the same point together with the opposite side of the triangle EFG (Art. [105]). The anharmonic ratio of any four points on a is equal to the anharmonic ratio of the four corresponding points on a' (Art. [117 (2)]). Hence, if six points on a form an involution, the corresponding six points on a' likewise form an involution.

It will be convenient to term conics which pass through the same three points, *three point conics*, and conics which pass through the same four points, *four point conics*.

The method may be illustrated as follows. Consider Desargues' theorem and let a', b', c' be three four point conics. Take three of the points of intersection of these conics as E, F, G. Then to a', b', c' correspond three straight lines a, b, c which meet at a point, since a', b', c' intersect in a fourth point. Hence since a', b', c' determine an involution on any transversal s', the corresponding points, viz. the points of intersection of a, b, c with the conic s, form an involution. Hence the theorem which corresponds to Desargues' theorem is to the effect that three concurrent straight lines determine an involution on a conic.

From the involution property of the quadrilateral it follows that if four three point conics intersect in pairs in A, A', B, B', C, C' the connectors of these points to any one of the three points of intersection of the conics form an involution pencil.

From Brianchon's theorem it follows that if six three point conics be described to touch a given straight line, three of the three point conics, which can be drawn through pairs of points of intersection of these six conics, meet at a point.

If a pair of the points E, F, G are the critical points the results obtained by this method are similar to those obtained by inversion.

Generally in this method (1) to a conic corresponds a quartic curve through E, F, G, having nodes at these points, (2) to a conic through one of the points E, F, G corresponds a cubic curve through E, F, G having a node at one of these points, (3) to a conic through two of the points E, F, G corresponds a conic passing through two of the vertices of the triangle, (4) to a conic through the three points E, F, G corresponds a straight line.

112. Harmonic locus of two imaginary conics.

As in Art. [133] it may be proved that *the envelope of a chord which is cut by two imaginary conics in pairs of points, which are harmonic conjugates, is a conic, which touches the eight tangents, which can be drawn to the two conics at their four points of intersection.* This conic is generally imaginary.

The correlative and also the converse theorems can also be proved as in Art, [133].

The envelope of chords cut harmonically by a conic and its conjugate imaginary curve is a real conic, which touches the eight tangents which can be drawn to these curves at their four points of intersection.

In this case the four points of intersection form a real or semi-real quadrangle $ABCD$.

(1) Let A, B, C, D be real. The tangents to the two conics at these points, being tangents to conjugate imaginary curves at the same real points on the curves, are pairs of conjugate imaginary lines. Thus, if a and a' be the tangents to the two conics at A, they are conjugate imaginary lines. Hence the harmonic locus of the conics touches four pairs of conjugate imaginary lines and is therefore real. (Art. 106.)

(2) Let A and B be real and C and D a pair of conjugate imaginary points. The tangents a and a' at A and the tangents b and b' at B to the two conics are two pairs of conjugate imaginary lines. Let c and c' be the tangents at C, and d and d' the tangents at D. Then as C is an imaginary point the lines c and c' are not conjugate imaginary lines. Consider however c and d'. They are tangents to conjugate imaginary curves at conjugate imaginary points and are therefore conjugate imaginary lines. Similarly c' and d are conjugate imaginary lines. Hence in this case also the harmonic locus touches four pairs of conjugate imaginary lines and is therefore a real conic. (Art. 106.)

(3) Let A and B and also C and D be pairs of conjugate imaginary points. Then as in the last case a and b', a' and b, c and d', and c' and d are pairs of conjugate imaginary lines and as they are tangents to the harmonic locus it is a real conic. (Art. 106.)

Hence the harmonic locus of two conjugate imaginary conics is a real conic. For its equation see Art. 114.

Similarly the correlative locus is a real conic.

113. Analytical.

Let $S=0$, $S'=0$, be the equations of two real conics, $\Sigma=0$, $\Sigma'=0$ their tangential equations, and $\Phi=0$ the tangential equation of the envelope of lines cut harmonically by the conics.

Consider the conic $S+iS'=0$. Its tangential equation is $\Sigma-\Sigma'+i\Phi=0$. Here the conic contains a real or semi-real quadrangle given by $S=0$ and $S'=0$, and has likewise a real or semi-real circumscribed quadrilateral whose sides are given by

$$\Sigma-\Sigma'=0, \quad \Phi=0.$$

The fact that $\Phi=0$ is one of these equations, shows that the real or conjugate imaginary tangents are tangents to the harmonic locus of $S=0$, and $S'=0$.

Consider the conic $\Sigma+i\Sigma'=0$. Its ordinary equation is $\Delta S-\Delta'S'+iF=0$, where $F=0$ is the locus of points from which tangents to $\Sigma=0$ and $\Sigma'=0$ form a harmonic pencil.

Hence the conic has a real or semi-real circumscribed quadrilateral the sides of which are given by $\Sigma=0$ and $\Sigma'=0$ and has likewise a real or semi-real inscribed quadrangle whose vertices are given by

$$\Delta S-\Delta'S'=0, \quad F=0.$$

The locus of points the tangents from which to the conics $\Sigma=0$ and $\Sigma'=0$ form a harmonic pencil passes through these points.

Let the equation of the conic be

$$S+iS'=0. \quad\dots\dots\dots\dots\dots\dots\dots\dots\dots\dots(a)$$

Consider the two real conics

$$S=0,$$

and

$$S'=0,$$

when these conics intersect in four real points or two pairs of conjugate imaginary points. The conics then have a real self-conjugate triangle. Hence S and S' may be written in the form :

$$S \equiv au^2+bv^2+cw^2,$$
$$S' \equiv a'u^2+b'v^2+c'w^2,$$

where u, v and w are linear functions of x and y.

Hence the equation (a) may be written

$$(a+ia')\,u^2+(b+ib')\,v^2+(c+ic')\,w^2=0,$$

or replacing u, v and w by x, y, z, as

$$(a+ia')\,x^2+(b+ib')\,y^2+(c+ic')\,z^2=0. \quad\dots\dots\dots\dots\dots(1)$$

The vertices of the inscribed quadrangle of this conic are given as the points of intersection of the real conics,

$$ax^2+by^2+cz^2=0, \quad\dots\dots\dots\dots\dots\dots\dots\dots(2)$$
$$a'x^2+b'y^2+c'z^2=0. \quad\dots\dots\dots\dots\dots\dots\dots\dots(3)$$

The tangential equation of (1) is

$$\frac{l^2}{a+ia'}+\frac{m^2}{b+ib'}+\frac{n^2}{c+ic'}=0, \quad\dots\dots\dots\dots\dots(4)$$

r $\quad\quad (a-ia')\dfrac{l^2}{a^2+a'^2}+(b-ib')\dfrac{m^2}{b^2+b'^2}+(c-ic')\dfrac{n^2}{c^2+c'^2}=0. \quad\dots\dots\dots(5)$

Hence the circumscribed quadrilateral is given by

$$a\,\frac{l^2}{a^2+a'^2}+b\,\frac{m^2}{b^2+b'^2}+c\,\frac{n^2}{c^2+c'^2}=0, \quad \dots\dots\dots\dots\dots(6)$$

and

$$a'\,\frac{l^2}{a^2+a'^2}+b'\,\frac{m^2}{b^2+b'^2}+c'\,\frac{n^2}{c^2+c'^2}=0. \quad \dots\dots\dots\dots(7)$$

(i) The values of x^2, y^2 and z^2 obtained from (2) and (3) are the same as the values of $\dfrac{l^2}{a^2+a'^2}$, $\dfrac{m^2}{b^2+b'^2}$, $\dfrac{n^2}{c^2+c'^2}$ obtained from (6) and (7). Hence, if the values of x, y, z are real, the values of l, m, n will be real. Hence, if the inscribed quadrangle is real, the circumscribed quadrilateral is real, and if the inscribed quadrangle is semi-real of the first kind the circumscribed quadrilateral is also semi-real of the first kind (cf. Art. 109).

(ii) Consider the polar reciprocal of (2) with regard to the conic

$$\sqrt{a^2+a'^2}\,x^2+\sqrt{b^2+b'^2}\,y^2+\sqrt{c^2+c'^2}\,z^2=0, \quad \dots\dots\dots\dots(8)$$

which is always real.

The line $lx+my+nz=0$ touches (2) if

$$\frac{l^2}{a}+\frac{m^2}{b}+\frac{n^2}{c}=0. \quad \dots\dots\dots\dots\dots\dots\dots(9)$$

The polar of x', y', z' with respect to (8) is

$$\sqrt{a^2+a'^2}\,xx'+\sqrt{b^2+b'^2}\,yy'+\sqrt{c^2+c'^2}\,zz'=0.$$

Substituting in (9) the locus of $x'y'z'$ is

$$(a^2+a'^2)\,\frac{x'^2}{a}+(b^2+b'^2)\,\frac{y'^2}{b}+(c^2+c'^2)\,\frac{z'^2}{c}=0.$$

The x, y, z equation of (6) is

$$(a^2+a'^2)\,\frac{x^2}{a}+(b^2+b'^2)\,\frac{y^2}{b}+(c^2+c'^2)\,\frac{z^2}{c}=0.$$

Hence the polar reciprocal of (2) with respect to (8) is (6).

Similarly   ,,   ,,   (3)   ,,   ,,   (8) is (7).

Hence the polars of the vertices of the quadrangle with respect to (8) are sides of the quadrilateral. This confirms (i).

(iii) The polar reciprocal of (1) with respect to (8) is

$$\frac{a^2+a'^2}{a+ia'}\,x^2+\frac{b^2+b'^2}{b+ib'}\,y^2+\frac{c^2+c'^2}{c+ic'}\,z^2=0,$$

or

$$(a-ia')\,x^2+(b-ib')\,y^2+(c-ic')\,z^2=0.$$

This is the conjugate imaginary curve of (1). Hence the polar reciprocal of (1) with respect to (8) is its conjugate imaginary curve.

114. Harmonic and anharmonic loci of two conics.

The equation of the envelope of a line, which is cut in constant anharmonic ratio by two conics $S=0$ and $S'=0$, can be easily obtained as follows :

All conics of the system $S+\lambda S'=0$ are cut by a transversal $lx+my+nz=0$ in an involution two pairs of conjugate points of which A, A' and B, B' are given as the points of intersection of the line $lx+my+nz=0$ with the conics $S=0$ and $S'=0$.

The double points of this involution are given by the values of λ, viz. λ_1 and λ_2, which are obtained from the tangential equation of $S+\lambda S'=0$, i.e. from

$$\Sigma + \lambda\Phi + \lambda^2\Sigma' = 0,$$

$$\therefore \lambda_1 + \lambda_2 = -\frac{\Phi}{\Sigma'},$$

and

$$\lambda_1\lambda_2 = \frac{\Sigma}{\Sigma'}.$$

But by example (6), Art. 27, if $(AA'BB')=\mu$,

$$4\left(\frac{1+\mu}{1-\mu}\right)^2 = \frac{(\lambda_1+\lambda_2)^2}{\lambda_1\lambda_2} = \frac{\Phi^2}{\Sigma\Sigma'}.$$

Therefore the equation of the anharmonic locus is

$$\Phi^2 = 4\left(\frac{1+\mu}{1-\mu}\right)^2 \Sigma\Sigma'.$$

The harmonic locus is $\Phi = 0$, for $1+\mu = 0$ in this case.

The equation of the common points is $\Phi^2 = 4\Sigma\Sigma'$, for $\mu = 0$ in this case.

Correlatively the equation of the locus of points, the tangents from which form a pencil of constant anharmonic ratio μ, is

$$F^2 = 4\left(\frac{1+\mu}{1-\mu}\right)^2 \Delta\Delta'SS'.$$

The corresponding harmonic locus is $F=0$, for $1+\mu = 0$ in this case.

The equation of the common tangents is $F^2 = 4\Delta\Delta'SS'$, for $\mu = 0$ in this case.

In the preceding

$$\Phi \equiv \Sigma\,(bc' + cb' - 2ff')\,l^2 \equiv \Sigma a_1 l^2 \text{ (suppose)},$$

and

$$F \equiv \Sigma\,(BC' + CB' - 2FF')\,x^2 \equiv \Sigma A_1 x^2 \text{ (suppose)}.$$

Consider the two conics

$$S + iS' = 0, \text{ and } S - iS' = 0. \dots\dots\dots\dots\dots\dots(1)$$

The tangential equations of these conics are

$$\Sigma - \Sigma' + i\Phi = 0 \text{ and } \Sigma - \Sigma' - i\Phi = 0.$$

The equations $\Phi = 0$ and $F = 0$ of the preceding become in this case

$$2\,(\Sigma + \Sigma') = 0, \dots\dots\dots\dots\dots\dots\dots\dots\dots(2)$$

and

$$2\,(\Delta S + \Delta'S' - F + \Phi_1) = 0,$$

where $\Phi_1 = 0$ is the $x,\ y,\ z$ equation of $\Phi = 0$.

Hence the equations of the anharmonic loci of the conics (1) are respectively

$$(\Sigma + \Sigma')^2 = \left(\frac{1+\mu}{1-\mu}\right)^2 \{(\Sigma - \Sigma')^2 + \Phi^2\}, \dots\dots\dots\dots(3)$$

and

$$(\Delta S + \Delta'S' - F + \Phi_1)^2 = \left(\frac{1+\mu}{1-\mu}\right)^2 \{(\Delta - \Theta')^2 + (\Theta - \Delta')^2\}(S^2 + S'^2). \dots\dots(4)$$

If $\mu=0$ (3) becomes $4\Sigma\Sigma'=\Phi^2$, which is the equation previously obtained for the points of intersection of the conics. '

The x, y, z equation of $2(\Sigma+\Sigma')=0$ is $2(\Delta S+\Delta'S'+F)=0$.

Consider the two conics
$$\Sigma+i\Sigma'=0 \text{ and } \Sigma-i\Sigma'=0.$$
The x, y, z equations of these conics are
$$\Delta S-\Delta'S'+iF=0 \text{ and } \Delta S-\Delta'S'-iF=0.$$

Their $\Phi=0$ and $F=0$ equations become respectively
$$2\frac{\Delta^2\Sigma+\Delta'^2\Sigma'-\Delta\Delta'\Phi+F_1}{(\Delta-\Theta')^2+(\Theta-\Delta')^2}=0,$$
$$2(S\Delta+S'\Delta')=0,$$
where $F_1=0$ is the tangential equation of $F=0$.

Hence their anharmonic loci are
$$(\Delta^2\Sigma+\Delta'^2\Sigma'-\Delta\Delta'\Phi+F_1)^2=\left(\frac{1+\mu}{1-\mu}\right)^2\{(\Delta-\Theta')^2+(\Theta-\Delta')^2\}^2(\Sigma^2+\Sigma'^2), \ldots\ldots(1)$$
$$(S\Delta+S'\Delta')^2=\left(\frac{1+\mu}{1-\mu}\right)^2\{(\Delta S-\Delta'S')^2+F^2\}. \ldots\ldots\ldots\ldots(2)$$

If $\mu=0$ equation (2) gives the same equation of the common tangents as that previously obtained.

Hence it follows that *the anharmonic loci of two conjugate imaginary conics are real conics.*

115. Anharmonic loci of two conjugate imaginary conics.

These may also be obtained as follows:

Consider the conics $S=0$ and $S'=0$.

The points of intersection of the line $lx+my+nz=0$ with the first of these are given by
$$(cl^2+an^2-2gln)x^2+2(clm-fln-gmn+hn^2)xy+(cm^2+bn^2-2fmn)y^2=0 \quad (1)$$
with a similar equation (2) for the points of intersection of the line with the second conic.

But if λ be one of the anharmonic ratios of the points given by
$$a_1x^2+2h_1x+b_1=0 \text{ and } a_1'x^2+2h_1'x+b_1'=0,$$
by example (3) Art. [14]
$$(2h_1h_1'-a_1b_1'-a_1'b_1)^2=4\left(\frac{1+\lambda}{1-\lambda}\right)^2(h_1^2-a_1b_1)(h_1'^2-a_1'b_1').$$

Now, if the values from (1) and (2) are substituted in this equation and Σ, Σ', Φ are the tangential equations of the conics and their harmonic locus, this becomes at once
$$\Phi^2=4\left(\frac{1+\lambda}{1-\lambda}\right)^2\Sigma\Sigma'. \ldots\ldots\ldots\ldots\ldots(3)$$

Consider the conics $S+iS'=0$ and $S-iS'=0$.

Their tangential equations are

$$\Sigma - \Sigma' + i\Phi = 0$$

and $$\Sigma - \Sigma' - i\Phi = 0,$$

and their harmonic locus is

$$\Sigma + \Sigma' = 0.$$

Hence from (3) their anharmonic locus is the real curve

$$(\Sigma+\Sigma')^2 = \left(\frac{1+\lambda}{1-\lambda}\right)^2 \{(\Sigma-\Sigma')^2+\Phi^2\}.$$

116. *The locus of real points at which a pair of conjugate imaginary points A and A' and a pair of imaginary points B and C subtend a pencil, whose anharmonic ratio is real, is a cubic curve.*

Let B' and C' be the conjugate imaginary points of B and C. Let P be a real point on the locus.

Then if $(P.AA'BC) = K + iK'$, $(P.A'AB'C') = K - iK'$.

Therefore since $K' = 0$,

$$(P.AA'BC) = (P.A'AB'C').$$

Hence the pencil subtended at P by the three pairs of points is an involution pencil and the required locus is the cubic obtained in Art. [142 (3)], which passes through the six points A, A', B, B', C, C'.

(1) If the chords AA', BB', CC' are concurrent at O the six points A, A', B, B', C, C' lie on a conic and the cubic breaks up into this conic and the polar of O with respect to the conic.   (See Art. [98].)

(2) In the general case the construction for certain points on the cubic is given in Art. [142 (3)].

In case (1) the points subtend at O a pencil whose anharmonic ratio may be 0, 1, or ∞, and in case (2) the same applies to points on the lines AA, BB', CC' in all cases, since two of the four rays coincide.

In the proof of the involution cubic given in Art. [143 (3)] the three pairs of points A, A ; B, B' ; C, C' are assumed to be real. From the preceding it follows that this need not be the case. Any of the pairs of points may be pairs of conjugate imaginary points. In case (1) the six points, whether real or conjugate imaginary, subtend an involution pencil at real points on the polar of O.

This result may be proved analytically as follows:

Take AA' for axis of x and let the coordinates of A and A' be ia' and $-ia'$. Let the coordinates of B and C be $b+ib'$, $c+ic'$ and $k+ik'$, $l+il'$ respectively.

Then the connectors of the points A, A', B, C to any point P (x, y) meet the axis of x in points given by

$$ia', \quad -ia', \quad \frac{(y-c)\,X-c'X'}{(y-c)^2+c'^2}+i\,\frac{(y-c)\,X'+c'X}{(y-c)^2+c'^2}, \quad \frac{(y-l)\,X_1-l'X_1'}{(y-l)^2+l'^2}+i\,\frac{(y-l)\,X_1'+l'X_1}{(y-l)^2+l'^2},$$

where $X=by-cx$, $X'=b'y-c'x$, $X_1=ky-lx$, $X_1'=k'y-l'x$.

The condition that the anharmonic ratio of a range should be real is given in Art. 24. The relation in this case can be reduced to the form

$$\begin{vmatrix} 1 & (y-c)^2+c'^2 & (y-l)^2+l'^2 \\ 0 & (y-c)\,X-c'X' & (y-l)\,X_1-l'X_1' \\ a'^2 & X^2+X'^2 & X_1^2+X_1'^2 \end{vmatrix}=0.$$

In this determinant, if y is made equal to zero, the result is zero. Hence y is a factor and the curve is a cubic.

117. *To find the equation of the conic, points on which subtend a pencil of anharmonic ratio $K+iK'$ at any four given points.*

Let the coordinates of the four given points A, B, C, D be x_1y_1; x_2y_2; x_3y_3; x_4y_4 respectively.

The connector of the point x_1y_1 to the point $P(xy)$ meets the axis of x in a point X_1, at a distance $\dfrac{xy_1-yx_1}{y_1-y}$ from the origin.

Substituting this and the similar values in the usual expression for an anharmonic ratio and equating the result to $K+iK'$, the equation of the conic is found to be

$$\frac{\begin{vmatrix} x & y & 1 \\ x_1 & y_1 & 1 \\ x_3 & y_3 & 1 \end{vmatrix} \begin{vmatrix} x & y & 1 \\ x_1 & y_1 & 1 \\ x_4 & y_4 & 1 \end{vmatrix}}{\begin{vmatrix} x & y & 1 \\ x_2 & y_2 & 1 \\ x_3 & y_3 & 1 \end{vmatrix} \begin{vmatrix} x & y & 1 \\ x_2 & y_2 & 1 \\ x_4 & y_4 & 1 \end{vmatrix}}=K+iK'.$$

If the four points A, B, C, D are real this result is at once obtained from the fact that

$$\sin APC=\frac{2}{AP\,.\,CP} \quad \text{area } APC=\frac{1}{AP\,.\,CP}\begin{vmatrix} 1 & 1 & 1 \\ x_1 & y_1 & 1 \\ x_3 & y_3 & 1 \end{vmatrix}$$

If A, B and C, D are pairs of conjugate imaginary points, while x and y are real, the expression on the left-hand side is real.

118. Theorem. *The anharmonic ratio of the four points of intersection of the conics* $S = 0$ *and* $S' = 0$ *at any point of the conic* $S - \lambda S' = 0$ *is* $\dfrac{K_1 - \lambda}{K_2 - \lambda} : \dfrac{K_1 - K_3}{K_2 - K_3}$, *where* K_1, K_2, K_3 *are the roots of the discriminating cubic of* $S - \lambda S' = 0$ *and the quantities* K_1, K_2, K_3 *and* λ *are interchanged according to the order in which the points of intersection of the conic are taken in the anharmonic ratio.*

Let $S - K_1 S' = 0$, $S - K_2 S' = 0$, $S - K_3 S' = 0$ determine respectively the pairs of lines $P_1 P_2$, $P_3 P_4$; $P_1 P_3$, $P_2 P_4$; $P_1 P_4$, $P_2 P_3$. Let O be any point in the plane and let $S - \lambda S' = 0$ be the conic through O, P_1, P_2, P_3, P_4. Then the equations of the polars of O with respect to the three pairs of lines and this conic are of the form $P - K_1 P' = 0$, $P - K_2 P' = 0$, $P - K_3 P' = 0$, $P - \lambda P' = 0$, the last being the equation of the tangent at O to the conic.

Now the polars of two fixed points with respect to four conics through four fixed points form two projective pencils. Hence the pencils formed by the polars of any two points with respect to the same four conics of the system have the same anharmonic ratios. But the polars of P_1 with respect to the three pairs of lines and the conic considered are $P_1 P_2$, $P_1 P_3$, $P_1 P_4$ and $P_1 P_1$, the tangent at P_1 to the conic $S - \lambda S' = 0$. But $(P_1 . P_2 P_3 P_4 P_1) = (O . P_2 P_3 P_4 P_1)$ by the anharmonic property of the conic. Hence $(O . P_2 P_3 P_4 P_1)$ equals the anharmonic ratio of the polars of O. That is to say it is equal to

$$\frac{K_2 - \lambda}{K_1 - \lambda} : \frac{K_2 - K_3}{K_1 - K_3}.$$

The author is indebted to Mr S. G. Soal for this proof. The result may also be proved analytically for the imaginary conic by means of Art. 117.

119. The theorem proved in the last article affords an easy method of distinguishing between real and imaginary conics which pass through the four vertices of a real or semi-real quadrangle.

Consider the conic $S - \lambda S' = 0$ and let the four points of intersection of the conics $S = 0$ and $S' = 0$ be real or two pairs of conjugate imaginary points. In this case K_1, K_2, K_3 are all real.

(i) Hence if the conic is real the anharmonic ratio of four real points or of two pairs of conjugate imaginary points on the curve is real: if the conic is imaginary the anharmonic ratio of the four real points on the curve or of the two pairs of conjugate imaginary points on the curve is imaginary.

Let the four points of intersection of the conics $S = 0$ and $S' = 0$ be two real points C and D and a pair of conjugate imaginary points A and A'. In this case

K_1 and K_2, two of the roots of the discriminating cubic, are a pair of conjugate imaginary quantities $a + ia'$, and $a - ia'$, while the third root K_3 is real.

Hence

$$(AA'CD) = \frac{a + ia' - K_3}{a - ia' - K_3} : \frac{a + ia' - \lambda}{a - ia' - \lambda} = K + iK' \text{ (suppose)}.$$

Let λ be real, then

$$(A'ACD) = \frac{1}{K + iK'}, = K - iK', \therefore K^2 + K'^2 = 1.$$

Let λ be "i," then $(A'ACD) \neq K - iK'$ and the relation $K^2 + K'^2 = 1$ does not hold.

(ii)   Hence if the conic is real an anharmonic ratio of two real points (C and D) and a pair of conjugate imaginary points (A, A') on the curve is $K + iK'$ where $K^2 + K'^2 = 1$; if the conic is imaginary the anharmonic ratio of the two real points (C, D) and the two conjugate imaginary points (A and A') on the curve is $K + iK'$ where $K^2 + K'^2 \neq 1$.

The only case in which the anharmonic ratio of the points of intersection of $S = 0$ and $S' = 0$ at any point of the imaginary conic $S + iS' = 0$ can be real is when two of the roots K_1, K_2, K_3 are equal, in which case the anharmonic ratio is 1, 0, or ∞.

If a conic passes through a real or semi-real quadrangle and through a real point, the curve is real, for the equation to determine λ does not involve "i."

If a conic passes through a real or semi-real quadrangle and through an imaginary point, the conic is real or imaginary according as it passes or does not pass through the conjugate imaginary point.   The condition necessary and sufficient is that the hexagon formed by the six points is a Pascal hexagon.   (See Art. 98.)

A real conic can always be described through three pairs of conjugate imaginary points the real connectors of which are concurrent, for the Pascal line of the hexagon so formed is the polar of the point of intersection of the connectors.

(iii)   Generally the anharmonic ratio of a pair of conjugate imaginary points (A, A') and of two imaginary points (B, C) on a real conic is imaginary.   If, however, the real chords through A, B, C are concurrent at a point O the anharmonic ratio is real.

Let B' and C' be imaginary conjugates of B and C.   Then if

$$(C' . AA'BC) = K + iK', \quad \langle C . A'AB'C') = K - iK'.$$

Since B' and C'' are on the conic

$$(AA'BC) = K + iK' \text{ and } (A'AB'C') = K - iK'.$$

If AA', BB', CC' are concurrent the six points form an involution on the conic, and $(AA'BC) = (A'AB'C')$.

$$K + iK' = K - iK'. \qquad K' = 0.$$

(iv)   Generally the anharmonic ratio of four imaginary points A, B, C, D on a real conic is imaginary.   If however the real chords through the points are concurrent at a point O the anharmonic ratio is real.

Let A', B', C', D' be the conjugate imaginary points of A, B, C, D.   They are on the curve.

Hence if $(ABCD) = K + iK', \quad (A'B'C'D') = K - iK'.$

But $(ABCD) = (A'B'C'D').$

Hence $K + iK' = K - iK', \quad \therefore \; K' = 0.$

In case (iii) when the chords are concurrent the four points subtend a pencil of an anharmonic ratio 0, 1, or ∞ at O, and a pencil of real anharmonic ratio at any real point on the polar of O. The latter follows from the fact that the six points A, A', B, B', C, C' subtend an involution at such a point (Art. 116).

In case (iv) when the chords are concurrent the four points subtend a pencil of real anharmonic ratio at O and also a pencil of real anharmonic ratio at any real point on the polar of O. The latter follows from the fact that the six points A, A', B, B', C, C' subtend an involution at such a point (Art. 116).

(v) Generally at any point on a real conic through the points the anharmonic ratio of

(1) Three real points and an imaginary point ;

(2) Two real points and two imaginary points ;

(3) One real point and three imaginary points ;

(4) One real point, a pair of conjugate imaginary points and an imaginary point

is imaginary.

In case (3) if the real lines through the three imaginary points (A, B, C) are concurrent at O and meet the conic again in A', B', C', their conjugate imaginary points, join O to D to meet the conic again in a real point D'. If D and D' are distinct the anharmonic ratio is imaginary, but if D and D' coincide, then

$$(ABCD) = (A'B'C'D').$$
$$K + iK' = K - iK'. \quad \therefore \; K' = 0.$$

Hence the anharmonic ratio is real.

Similarly the anharmonic ratio in case (4) may under certain circumstances be real.

120. Construction of conics from real data.

If a conic is given by the fact that it either passes through real points or touches real straight lines, it is easily possible to distinguish between the cases when this conic is real and when it is imaginary. Using the notation of Art. [144] and denoting a real conic by 1 and an imaginary conic by 1.i, the results are as follows :

(1) $[:\cdot:] = 1,$ (4) $[:///] = 4$ or $4.\,i,$

(2) $[::/] = 2$ or $2.\,i,$ (5) $[\cdot////] = 2$ or $2.\,i,$

(3) $[\cdot\cdot//] = 4$ or $4.\,i,$ (6) $[/////] = 1.$

In case (2) the conic is real or imaginary according as the double points of the involution determined by the four points on the line are real or imaginary.

If in any position of the line the double points are real and the line be rotated round a fixed point on it, till it passes through one of the four fixed points, A, the double points in the new position coincide with A and the two conics are coincident. After the line has passed through A, the double points become imaginary and the two conics are a pair of conjugate imaginary conics.

In case (3) the four conics are either all real or are two pairs of conjugate imaginary conics. By Art. [113 (E)] the points of contact of the conics with the two straight lines are the intersections of these lines with the connectors of two pairs of double points of two involutions situated on real lines. If the four double points are real the intersections of their connectors with the given lines are real and therefore the four conics are real. If one pair of these double points (which form a real or semi-real quadrangle) or both pairs, are pairs of conjugate imaginary. points, the points of contact of the conics with the two lines are imaginary and the conics are two pairs of conjugate imaginary conics.

This may be proved as follows. Let the given lines be p and p' and the two pairs of double points conjugate imaginary points A, A' and B, B'. Then AB and $A'B'$ are a pair of conjugate imaginary lines and they meet p and p' in two pairs of conjugate imaginary points. Hence one pair of the given conics are a pair of conjugate imaginary conics. Similarly, since AB' and BA' are a pair of conjugate imaginary lines, the other pair of conics are a pair of conjugate imaginary conics.

If A and A' be a pair of real points, say D and E, and B and B' be a pair of conjugate imaginary points, the lines DB, DB' are a pair of conjugate imaginary lines as are also the lines EB and EB'. Hence the same result follows in this case.

121. In the statement Art. 120 a pair of conjugate imaginary points may be substituted for a pair of real points, or a pair of conjugate imaginary lines for a pair of real lines, in the determining elements (see Art. 106), except (a) that in (2), if the four points are two pairs of conjugate imaginary points, the conics must be real (Art. 19, case 11) with a similar alteration in (5); and (b) that in (3), if the three points are real and the tangents are conjugate imaginary lines, the conics are real as the double points of the involutions are real, and, if of the three points two are conjugate imaginary, then a pair of the conics are real and a pair are conjugate imaginary conics, with similar alterations in (4). [See Art. 106.]

122. Foci of an imaginary conic.

If the equation of the conic be

$$ax^2 + 2hxy + by^2 + 2gx + 2fy + c = 0,$$

it can be easily proved that the coordinates of the foci (a, β) are given by the equations

$$a^2 = \beta^2 = (a - b)\frac{\Delta}{C^2} \quad \text{and} \quad a\beta = h\frac{\Delta}{C^2},$$

after the origin has been transferred to the point $\frac{G}{C}, \frac{F}{C}$.

It can also be proved that, if T be the sum of the measures of the distances of any point on the curve from a pair of conjugate foci,

$$T^2 = 2\{\pm \sqrt{(a-b)^2 + 4h^2} - (a+b)\}\frac{\Delta}{C^2},$$

where the $\pm$ sign is to be taken according to the pair of conjugate foci considered.

Hence the sum of the measures of any point on an imaginary conic from a pair of conjugate foci is constant.

123. Every imaginary conic contains a real or semi-real quadrangle which by a real projection (Art. 86) may be projected into a square $ABCD$, real, semi-real of the first kind, or semi-real of the second kind. Take the real diagonals of this square, OA and OB, as axes of x and y respectively.

Then in the three cases the coordinates of the vertices are respectively:

	A	B	C	D
1st case	$a, 0$	$0, a$	$-a, 0$	$0, -a$
2nd case	$ia, 0$	$0, ia$	$-ia, 0$	$0, -ia$
3rd case	$a, 0$	$0, ia$	$-a, 0$	$0, ia$

Hence if λ be one of the anharmonic ratios, which any point on the curve subtends at these points, the equations of the conic in the three cases are from Art. 117, respectively:

$$(1) \qquad x^2 + y^2 - a^2 = 2xy\frac{\lambda+1}{\lambda-1},$$

$$(2) \qquad x^2 + y^2 + a^2 = 2xy\frac{\lambda+1}{\lambda-1},$$

$$(3) \qquad -x^2 + y^2 + a^2 = 2ixy\frac{\lambda+1}{\lambda-1}. \quad \text{Cf. Art. 137.}$$

124. Foci of the conics into which a pair of conjugate imaginary conics can be projected.

In cases (1) and (2) of the last article the conic and its conjugate imaginary conic have their axes in the same real directions and have a common real centre (Art. 86). It can be proved geometrically or analytically that the eight foci of the two conics are four pairs of conjugate imaginary points, which lie four on each of the two real axes.

If Ω_1 and Ω_2 be the critical points and the tangents from Ω_1 to the two conics are a_1, a_1', and b_1, b_1', and from Ω_2 to the two conics a_2, a_2' and b_2, b_2', then the pairs of tangents which are conjugate imaginary lines may be taken as a_1 and b_2, a_1' and b_2', a_2 and b_1, a_2' and b_1'. Consider the quadrilateral $a_1a_1'b_2b_2'$. The points a_1b_2 and $a_1'b_2'$ are real points, also a_1a_1' gives the conjugate imaginary point to b_2b_2'. Hence the two diagonals of the quadrilateral, which are at a finite distance, are real. They may be easily shown to be at right angles, and their point of intersection is the centre of the two conics. There is a second quadrilateral $a_2a_2'b_1b_1'$ with similar properties.

125. Real conics having double contact at a pair of conjugate imaginary points.

Consider the real conics $S=0$ and $S-\lambda a^2=0$, where a is a linear expression in x and y.

If the line $a=0$ meets the conic $S=0$ in a pair of conjugate imaginary points the conics have double contact at these points.

Take a line through the centre parallel to $a=0$ as axis of y and its conjugate diameter with respect to $S=0$ for axis of x. The equations of the two conics will then be of the form

$$ax^2+by^2-1=0,$$

and
$$ax^2+by^2-1-k(x-l)^2=0.$$

These conics can be graphed for the pair of conjugate diameters, which are the axes of x and y, and it will be found that their imaginary branches touch.

The properties of conics having double contact at real points, which were proved in Art. [130], are true also for conics having double contact at pairs of conjugate imaginary points.

In Art. [134] it was proved that the harmonic locus of a conic and a pair of points was related to the conic in such a way that if a simple quadrilateral was circumscribed to the conic and inscribed in the harmonic locus, an infinite number of quadrilaterals could be similarly described.

If the pair of points are on the conic, the harmonic locus of these points and the conic has double contact with the conic at these points. Hence given a real conic and a real line, another conic can be described to have double contact with the first conic at the points where it is met by the line, which also possesses the property that an infinite number of simple quadrilaterals can be circumscribed to the first conic and inscribed in the latter.

This should be compared with the property of conics having double contact proved in Art. [131]. The conics in question possess the remarkable property that not only do the sides of the inscribed quadrilateral meet in two points on the chord of contact, but the connectors of their points of contact with the other conic also pass through these same points.

The properties of conics having double contact at a pair of conjugate imaginary points may be obtained at once from the fact that, in a real plane perspective, two such conics may be made to correspond to a pair of concentric circles.

CHAPTER VI

TRACING OF CONICS AND STRAIGHT LINES

126. Graphic representation of the imaginary.

Although in the preceding pages of this book the term "graph of an imaginary point" has been used, and in certain places—especially in Chapter II—diagrams of imaginary branches of curves have been given, such use of "graphs" and figures is not essential, and these have been used rather as a convenient way of expressing what was meant or of explaining and illustrating results, than because their use was essential.

The graphical representation of the imaginary must always, from the nature of the original hypothesis, be difficult and defective. Still some representation is felt to be better than none—as long as the limitations of the method employed are clearly understood. In this chapter, for this reason, the representation of imaginary and complex points by means of what were termed in Chapter II Poncelet figures will be more fully considered.

In Chapter I, for convenience the graph of an imaginary point on a fixed straight line at a distance il from a fixed point O on the line was defined as being the point at a distance l from the point O. Hence, to obtain a representation of a point P, whose position is determined by an imaginary length ix' measured parallel to a real axis of x and a real length y' measured parallel to a real axis of y, it is natural to measure a length ON along OX equal to x' and a length NP' equal to y' and parallel to OY, and thereby to obtain a point P', which represents P graphically.

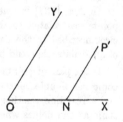

A point Q, referred to the same axes of coordinates and determined by coordinates x'', iy'', may be represented in the same way, remembering that in this case distances parallel to the axis of y are imaginary and those parallel to the axis of x are real. This is possible, but there are certain advantages in modifying somewhat this system of representation.

In this modification—the advantages of which have already been apparent in Arts. 72 and 77—with the usual system of real axes, the positive imaginary axis of x is taken along OX and the negative imaginary axis of x along OX', but the positive imaginary axis of y

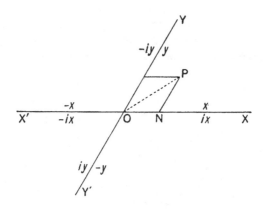

is taken along OY' and the negative imaginary axis of y along OY. This system of coordinates will be adopted generally in this chapter.

Representation of imaginary straight lines through the origin.

Consider the straight line $ax + iby = 0$. Here $\dfrac{x}{-iy} = \dfrac{b}{a}$ Hence the line may be represented by OP in the figure. But the equation of the line may be written $iax - by = 0$. Hence $\dfrac{ix}{y} = \dfrac{b}{a}$ Therefore in this case also the line is represented in the figure by OP. Similarly the line $ax - iby = 0$ has the same graph in whichever way its equation is written.

127. Tracing of real conics.

If imaginary and complex points are taken into account in tracing a conic, $x_1 + ix_2$ and $y_1 + iy_2$ may be substituted in the equation of the curve for x and y, where x_1, x_2, y_1, y_2 are real. If real and imaginary parts of the resulting equation are equated to zero, two equations connecting the four quantities x_1, x_2, y_1, y_2 are obtained. Arbitrary values may be given to any two of these four quantities and the values of the remaining two may be obtained from the equations. Hence it follows that the equation represents graphically not a single curve but a system of curves.

(a)   To trace the curve

$$\frac{x^2}{a^2} + \frac{y^2}{b^2} = 1.$$

All points on the curve are obtained by giving to x the value $x_1 + ix_2$, where x_1 and x_2 can have all real values from $-\infty$ to $+\infty$. No two points thus obtained can coincide, for their x coordinates will in all cases be different.

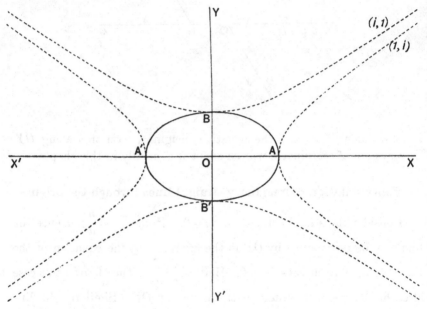

Let x_2 be zero.

(1)   Give x_1 all values from $-a$ to $+a$. Real values of y are then obtained and the part of the curve given by the continuous line is the corresponding branch.   This may be termed the (1, 1) branch.

(2)   Give x_1 all values from $-\infty$ to $-a$ and from $+a$ to $+\infty$. Conjugate imaginary values of y are obtained and the dotted part of the figure marked (1, i) is the corresponding branch.   This is the real part of the curve $\frac{x^2}{a^2} - \frac{y^2}{b^2} = 1$.   For this part of the curve the coordinates are x and iy.   It may be termed the (1, i) branch.

Let x_1 be zero.

(3)   There are no points in this case, the coordinates of which are

of the form iy, but such a branch may exist. When it exists, it may be termed the (i, i) branch.

(4) Give x_2 all values from $-\infty$ to $+\infty$. The corresponding values of y are real and the corresponding branch of the curve is that marked $(i, 1)$ in the figure. For this part of the curve the coordinates are ix and y. It is the real part of the curve $\dfrac{y^2}{b^2} - \dfrac{x^2}{a^2} = 1$. It may be termed the $(i, 1)$ branch.

The above may be called the *parent* curve, which consists of a real or $(1, 1)$ branch and a purely imaginary or (i, i) branch, together with a $(1, i)$ branch and an $(i, 1)$ branch.

It would be possible to determine further points on the curve by giving a constant value to x_2. In this case it would be necessary to find the corresponding values of y for all real values of x_1 in the equation

$$\frac{(x_1 + ix_2)^2}{a^2} + \frac{y^2}{b^2} = 1,$$

where x_2 is constant. This is the same conic as $\dfrac{x^2}{a^2} + \dfrac{y^2}{b^2} - 1 = 0$ with its centre displaced along the axis of x through a negative distance ix_2. Such a curve however is an imaginary conic, and conjugate imaginary points are not as a general rule situated on the locus. Hence it is more satisfactory to find the further imaginary points on the curve by another method.

Two conjugate imaginary points $x_1 + ix_2$, $y_1 + iy_2$ and $x_1 - ix_2$, $y_1 - iy_2$ lie on the real straight line $\dfrac{x - x_1}{x_2} = \dfrac{y - y_1}{y_2}$. Their mean point is the real point x_1, y_1, which is on this straight line. The equation of the tangent at either of the points, where the line joining the origin to the point x_1, y_1 meets the curve, is of the form

$$\frac{\lambda x_1 x}{a^2} + \frac{\lambda y_1 y}{b^2} = 1.$$

This is parallel to the line joining the conjugate imaginary points if $\dfrac{x_1 x_2}{a^2} + \dfrac{y_1 y_2}{b^2} = 0$. This relation holds if the pair of conjugate imaginary points lie on the curve (see Art. 130). Hence the real line joining the pair of conjugate imaginary points and the connector of their mean point to the origin are parallel to a pair of conjugate diameters of the conic. If these conjugate diameters are taken as axes of coordinates, the equation of the conic is

$$\frac{x^2}{a'^2} + \frac{y^2}{b'^2} - 1 = 0,$$

where a' and b' are the lengths of the corresponding semi-conjugate diameters.

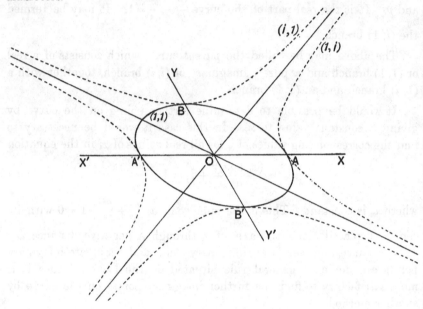

If this curve be traced for real and for purely imaginary values of the coordinates, as the curve $\frac{x^2}{a^2} + \frac{y^2}{b^2} - 1 = 0$ was traced, certain new branches of the curve are found. Thus:

(1) the (1, 1) branch is found as before,

(2) the $(1, i)$ branch is replaced by another hyperbola, which touches the real branch at the ends of the diameter, which is the axis of x,

(3) the (i, i) branch in this case, as before, does not exist,

(4) the $(i, 1)$ branch is replaced by another hyperbola, which touches the real branch at the ends of the diameter, which is the axis of y.

By taking different pairs of conjugate diameters for axes of coordinates all points on the conic are obtained and no points except those on the real and the purely imaginary branch occur more than once. These branches may be called Poncelet figures. They are also, for reasons given hereafter, termed (α, β) branches of the curve, where α and β are the angles which the pair of conjugate diameters, which determine them, make with the major or transverse axis of the conic.

(b) To trace the curve

$$\frac{x^2}{a^2} - \frac{y^2}{b^2} = 1.$$

It will be noticed that this equation may be written

$$\frac{x^2}{a^2} + \frac{(iy)^2}{b^2} = 1.$$

Hence the same graph is obtained as in the previous case, provided the axis of y be changed from real to imaginary and vice versâ.

Thus in this case:

(1) gives an ellipse, as in the previous figure, which is the $(1, i)$ branch,

(2) gives a real hyperbola as in the previous figure, which is the $(1, 1)$ branch,

(3) this branch, which is the $(i, 1)$ branch, again does not exist,

(4) gives the same curve as before, but it is a. purely imaginary (i, i) curve.

(c) To trace the curve

$$-\frac{x^2}{a^2} + \frac{y^2}{b^2} = 1.$$

This may be written

$$\frac{(ix)^2}{a^2} + \frac{y^2}{b^2} = 1.$$

In this case:

(1) gives an ellipse of the form $(i, 1)$,

(2) gives a hyperbola of the form (i, i),

(3) does not exist, but if it did it would be of the form $(1, i)$,

(4) gives a hyperbola of the form $(1, 1)$, i.e. a real curve.

In cases (b) and (c) the (α, β) branches may be obtained in the same way as for the curve $\frac{x^2}{a^2} + \frac{y^2}{b^2} - 1 = 0$.

(d) To trace the curve

$$\frac{x^2}{a^2} + \frac{y^2}{b^2} + 1 = 0.$$

This equation may be written

$$\frac{(ix)^2}{a^2} + \frac{(iy)^2}{b^2} - 1 = 0.$$

Hence:

(1) gives an ellipse of the form $(i. \ i)$,

(2) gives a hyperbola of the form $(i, 1)$,

(3) does not exist, but if it did it would be of the form $(1, 1)$,

(4) gives a hyperbola of the form $(1, i)$.

It follows that all central conics, when traced for all real, imaginary and complex points on the curve, have figures of the same type, and that a real ellipse, an imaginary ellipse and a hyperbola differ from each other only in so far that the $(1, 1)$, (i, i), $(1, i)$ and $(i, 1)$ branches are interchanged and that different branches of the parent curve develope into the (α, β) branches. Thus in the case of a real ellipse the (α, β) branches are hyperbolae, and in the case of a real hyperbola the (α, β) branches are ellipses.

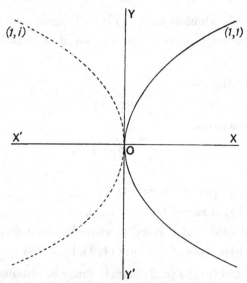

(e) To trace the curve

$$y^2 - 4ax = 0.$$

In this case it will be more convenient to treat y as the independent variable and to assume that it is of the form $y_1 + iy_2$, where y_1 and y_2 may have any real values.

Let y_2 be zero.

(1) If y_1 be given all real values from $-\infty$ to $+\infty$, the corresponding values of x are real and positive, and the real parabola marked with a continuous line in the figure is obtained.

Let y_1 be zero.

(2) If y_2 be given all real values from $-\infty$ to $+\infty$, the corresponding values of x are real and negative, and the parabola marked with a dotted line in the figure is the curve obtained. This is a parabola equal to the real parabola but with its axis turned in the opposite direction. This is the $(1, i)$ branch.

The branches (1) and (2), which are the $(1, 1)$ and $(1, i)$ branches, constitute the parent curve.

The (α, β) loci are obtained by taking any diameter and the tangent at the point, where it meets the real branch, as axes of coordinates. They consist of parabolas with axes parallel to the chief axis of the parabola and all touch the real branch. They are obtained by tracing the equation of the parabola in the form $y^2 - 4a'x = 0$.

128. Special case of the real circle.

The locus represented by $x^2 + y^2 + a^2 = 0$ is according to definition a real conic, although it has no real branch.

It has an (i, i) branch, the circle in the figure, and also an $(i, 1)$ branch and a $(1, i)$ branch, both of which are rectangular hyperbolae.

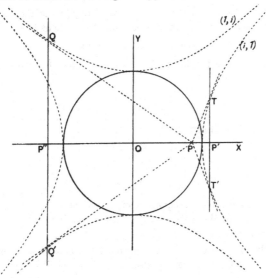

Any real line through the centre O meets the curve in a pair of conjugate imaginary points, which are the double points of a real involution. Pairs of conjugate points are, if real, on different sides of O and are such that $OP \cdot OP'' = -a^2$. If the distance of one point from O is a purely imaginary quantity, the distance of the other is purely imaginary and the points are on the same side of O. Pairs of conjugate points of this involution are inverse points with respect to the curve.

If P be taken such that the distance OP is purely imaginary, the polar of P is the line $TP'T'$ in the figure.

If P be taken such that the distance OP is real, the polar of P is the line $QP''Q'$ in the figure.

Since the coordinate geometry of purely imaginary points is the same as that of real points, the properties of purely imaginary points on the curve are the same as those of real points on the real circle.

Special case of the real conic.

The locus represented by $\frac{x^2}{a^2} + \frac{y^2}{b^2} + 1 = 0$ is according to definition a real conic.

Its properties may be deduced from those of the real ellipse $\frac{x^2}{a^2} + \frac{y^2}{b^2} - 1 = 0$ in the same way that the properties of the curve $x^2 + y^2 + a^2 = 0$ can be deduced from those of the circle $x^2 + y^2 - a^2 = 0$. This conic has a director circle the equation of which is $x^2 + y^2 + a^2 + b^2 = 0$ and likewise a pair of real foci situated on the minor axis. It may be shown that the sum of the measures of the distances of these points from any point on the curve is equal to $2ib$. The corresponding directrices are real.

129.   The fact that there is only one tangent at any point to a conic may be verified as follows.

Let $y = f(x)$ be the equation of a curve so that $f(x_1)$ is the value of y corresponding to a value x_1 of x. Then, if h be a small increment to x_1,

$$f(x_1 + h) = f(x_1) + hf'(x_1) + \dots.$$

The connector of the points given by x_1 and $x_1 + h$, disregarding powers of h above the first, is

$$\begin{vmatrix} X & Y & 1 \\ x_1 & f(x_1) & 1 \\ x_1 + h & f(x_1) + hf'(x_1) & 1 \end{vmatrix} = 0,$$

or

$$\begin{vmatrix} X & Y & 1 \\ x_1 & f(x_1) & 1 \\ 1 & f'(x_1) & 0 \end{vmatrix} = 0.$$

As this does not involve h, the equation of the connector of the points does not depend on the nature of the increment given to x_1, i.e. the same result is obtained whether the increment is real or imaginary.

Geometrically the result seems to depend on the fact that infinitely small and infinitely large quantities may be regarded in the limit as either real or imaginary.

130. *To find the condition that the point* $x_1 + ix_2$, $y_1 + iy_2$, *which is* *on the curve* $\dfrac{x^2}{a^2} + \dfrac{y^2}{b^2} = 1$, *should lie on that branch of the curve whose* *axes are the conjugate diameters, which are inclined at angles* α *and* β *to* *the axis major of the conic.*

If x and iy are the coordinates of the point referred to the conjugate diameters, given by α and β, as axes, x is equivalent to $x \cos \alpha$ along the axis of x, and $x \sin \alpha$ along the axis of y, while iy is equivalent to $iy \cos \beta$ along the axis of x, and $iy \sin \beta$ along the axis of y.

Hence the coordinates of the point x, iy, referred to the principal axes of the conic are

$$x_1 + ix_2 = x \cos \alpha + iy \cos \beta, \quad y_1 + iy_2 = x \sin \alpha + iy \sin \beta.$$

Therefore

$$x_1 = x \cos \alpha, \quad y_1 = x \sin \alpha, \quad x_2 = y \cos \beta, \quad y_2 = y \sin \beta.$$

Therefore $\quad \dfrac{y_1}{x_1} = \tan \alpha, \quad \dfrac{y_2}{x_2} = \tan \beta.$

Hence the coordinates of a point on the (α, β) branch are of the form

$$x_1 + ix_2, \quad x_1 \tan \alpha + ix_2 \tan \beta.$$

If the point $x_1 + ix_2$, $y_1 + iy_2$ is on the conic

$$\frac{x_1^2 - x_2^2}{a^2} + \frac{y_1^2 - y_2^2}{b^2} = 1 \quad \text{and} \quad \frac{x_1 x_2}{a^2} + \frac{y_1 y_2}{b^2} = 0.$$

Therefore $\quad \dfrac{y_1 y_2}{x_1 x_2} = -\dfrac{b^2}{a^2} = \tan \alpha \tan \beta.$

If points, whose coordinates are $x_1 + ix_2$, $x_1 \tan \alpha + ix_2 \tan \beta$ and $x_1' + ix_2'$, $x_1' \tan \alpha + ix_2' \tan \beta$, are on the conic, they lie on the same branch of it. For different values of x_1, x_2, α and β, the point having coordinates of this form may be any point. Hence by substituting these expressions in the equation of a central conic the points on the different (α, β) figures are obtained.

It will be noticed that, for a point to be on the curve,

$$\tan \alpha \tan \beta = -\frac{b^2}{a^2},$$

and that if α and β are connected by this relation, there are an infinite number of points on the corresponding branch of the curve.

If the relation $\tan \alpha \tan \beta = -\dfrac{b^2}{a^2}$ is not satisfied, there are no points on the curve. Thus, corresponding to a diameter given by α, there are only imaginary points on the curve when the conjugate diameter is associated with it. The planes or figures obtained by associating together the values of α and β, which determine conjugate diameters of the conic, will be termed the *nest of the conic*. In this respect the real conic is essentially different from the imaginary conic for which there is no nest. The general use of coordinates of the form $x_1 + ix_2$ and $x_1 \tan \alpha + ix_2 \tan \beta$ is considered in the following Articles.

131. General conclusions.

In the parent branch of a real conic, it has been shown that when the graph is formed for the axes of the conic, there are or may be four parts, viz.

(1) The real or (1, 1) branch.

(2) The purely imaginary or (i, i) branch.

(3) The $(1, i)$ branch which is an Argand diagram, the real axis being the axis of x.

(4) The $(i, 1)$ branch which is an Argand diagram, the real axis being the axis of y.

The (1, 1) branch represents the curve as usually considered.

The (i, i) branch is the curve, which the equation represents on the assumption that the squares of all lines are negative. With this assumption it is as real as the real branch and it has identical properties.

The branches $(1, i)$ and $(i, 1)$ differ in one respect from the Argand diagram in common use. In an Argand diagram it is usual to assume that, if the real variable and also the imaginary variable are infinite, *the* point at infinity is obtained. This is not the case in the present instance. There are at infinity at any rate two points—if not four— one given by an infinite positive value of the real variable and an infinite positive value of the imaginary variable, the other by an infinite positive value of the real variable and an infinite negative value of the imaginary variable. These two points are the circular points at infinity which from a certain point of view are more distant from the origin than other points on the line joining them, which is the line at infinity. An Argand diagram is of course a graphic representation of the quantities which make up a complex variable.

If the conic be graphed for a pair of conjugate diameters making angles α and β respectively with the axis major of the conic, it has been shown that there are again four branches, viz.

(1) The real or (1, 1) branch.

(2)′ The purely imaginary or (i, i) branch.

(3) The $(1, i)$ branch, in which the axis of x is the real axis.

(4) The $(i, 1)$ branch, in which the axis of y is the real axis.

The branches (1, 1) and (i, i) are the same graphically as those of the parent graph, but the branches (3) and (4) are different and vary with different values of α and β. They may be regarded as Argand diagrams in which the axes of coordinates are not at right angles.

Whatever value is given to α there is always one corresponding value of β, and for every value of β there is one value of α. The graph gives the point of intersection with the curve of real lines parallel to the β axis. Hence with a series of such figures the points of intersection of all real lines with the conic are given graphically. Also since one real line passes through every imaginary point, all points on the curve are graphically represented.

132. General representation of points by Poncelet or (α, β) figures.

The preceding renders it possible to conceive the points which exist in a plane, when the values of the determining coordinates of a point are or may be complex.

Through any point, which may be taken as origin, draw two rectangular axes. There are with respect to these rectangular axes four systems of points, viz.

(1) Those whose coordinates are real or (1, 1) points.

(2) „ „ „ purely imaginary or (i, i) points.

(3) „ „ „ of the form $(1, i)$.

(4) „ „ „ of the form $(i, 1)$.

Take any real line through the origin making an angle α with the axis of x. With this may be associated any other real line through the origin making an angle β with the axis of x. Let lengths real and purely imaginary be measured along or parallel to these lines, and let such lengths be regarded as the coordinates of a point.

Then, as in the preceding case, there are four systems of points, viz.

(1) Those whose coordinates are real or (1, 1) points.

(2) „ „ „ purely imaginary or (i, i) points.

(3) „ „ „ of the form $(1, i)$.

(4) „ „ „ of the form $(i, 1)$.

The $(1, 1)$ and (i, i) points are the same as in the preceding case. The $(1, i)$ and $(i, 1)$ points are different. By varying α and β the coordinates of all complex points in the plane may be thrown into this form.

Any value of β may be associated with any value of α. Hence since the graphs of a real conic are such that $\tan \alpha \tan \beta = -\dfrac{a}{b}$, where $ax^2 + by^2 - 1 = 0$ is the equation of the curve, it follows that the graphs of a real conic lie in a limited number of the (α, β) planes, and in each plane in which the graph exists there are an infinite number of points.

The same is true for a pair of conjugate imaginary lines when they are graphed with respect to their real point as origin. (See Art. 76.)

The coordinates of a point as set forth in the preceding have been termed the principal coordinates of the point (Art. 8). To find the points on a given line which are in an (α, β) plane, it is only necessary to substitute in the equation the values $x_1 + ix_2$, $x_1 \tan \alpha + ix_2 \tan \beta$, and equating real and imaginary points to find the corresponding real values of x_1 and x_2. It is sometimes more convenient to substitute $x_1 + iy_2 \tan \gamma$ and $x_1 \tan \alpha + iy_2$ for x and y.

In order to obtain graphically the points of intersection of two curves, it is not necessary that the α and β of the two graphs should be the same. It is only necessary that the β's should have the same value, when the two origins are real with respect to each other. Thus in Art. 49 when the points of intersection of a conic, having a real branch, with a pair of conjugate imaginary lines were obtained, the direction of β was obtained from geometrical considerations. The corresponding values of α for the two graphs were deduced and the points of intersection of the curves were obtained. This is always possible when the two origins are real with reference to each other. If the origins are purely imaginary points with reference to each other it is necessary for the real axis to have the same direction in both graphs. Generally in the case of the intersections of an imaginary line with a conic—having a real branch—the points of intersection lie in two different (α, β) figures, but, if the two values of β are equal, they lie in the same (α, β) figure. Such a case arises, if the line is the connector of a pair of points on the curve which lie in the same (α, β) figure.

133. Change of origin in the case of graphs.

(*a*) A curve may be graphed as already set forth for (α, β) planes with respect to some centre of symmetry such as the centre of a conic

or the real point on an imaginary line. Then, if the origin be transferred
to any real point, the graph will be the same but the α will be changed
and will not be constant for points on the same branch. This arises
from the fact that the real coordinate of each point on the branch
considered has to be combined with the real coordinates of the new
origin. Hence the direction of the real coordinate is changed and
the α becomes different for different points on the same (α, β) figure of
the original origin. The β however is unaltered.

(*b*) A similar process may be employed, when the new origin is a
purely imaginary point. In this case however the α remains unaltered
for points in the same (α, β) plane, while the β is changed.

(*c*) If the new origin is a complex point with respect to the original
axes, its position may be determined by means of its principal coordinates.
In this case a new real length and a new imaginary length have to
be combined with the α and β lengths of the graphs of each point so
that the α and β of each point of the graph are changed.

From the above it is seen how the vector method of treating
imaginary coordinates becomes possible in the case of the general
theory.

134. Poncelet or (a, β) figures.

In each case $x_1 + ix_2$ and $x_1 \tan a + ix_2 \tan \beta$ are substituted for x and y in the
equation of the locus, and the real and imaginary parts of the resulting equation are
equated to zero.

(*a*) *Real straight line.*

(1) *Origin on the line.*

Let the equation of the line be $ax + by = 0$. The resulting equations are

$$x_1 (a + b \tan a) = 0, \quad x_2 (a + b \tan \beta) = 0.$$

For these to be satisfied by values of x_1 and x_2 other than zero, it is found that

$$\tan a = \tan \beta = -\frac{a}{b}.$$

Hence it follows that the real and imaginary axes must coincide with the real
line, and therefore the real line gives the full graphic representation of the line (cf.
Art. 1).

(2) *Origin any real point.*

Let the equation of the line be $ax + by + c = 0$. The resulting equations are

$$x_1 (a + b \tan a) + c = 0, \quad x_2 (a + b \tan \beta) = 0.$$

Hence $\tan \beta = -\frac{a}{b}$. Substituting this value in the first equation, it follows that

$$x_1 (\tan a - \tan \beta) + \frac{c}{b} = 0.$$

Hence, if $\tan \beta = -\dfrac{a}{b}$, there are for each value of a an infinite number of values of x_2 but one definite value of x_1. Hence the points all lie in the plane (1, 1) and the planes $\left(a, \tan^{-1} \left(-\dfrac{a}{b} \right) \right)$. This is geometrically consistent with the previous result.

(b) *Real conic.*

(1) *Origin the centre.*

Let the equation of the conic be $ax^2 + 2hxy + by^2 = 1$. The resulting equations are

$$x_1^2 \{a + 2h \tan a + b \tan^2 a\} - x_2^2 \{a + 2h \tan \beta + b \tan^2 \beta\} = 1$$

and $\qquad x_1 x_2 \{a + h (\tan a + \tan \beta) + b \tan a \tan \beta\} = 0.$

Hence, if $a + h (\tan a + \tan \beta) + b \tan a \tan \beta = 0$, there are an infinite number of points in the plane (a, β). That is, there are an infinite number of points in the planes of the nest of the conic but in no other (a, β) planes.

(2) *Origin any real point.*

Let the equation of the conic be $f(x, y) = 0$. The resulting equations are

$$f(x_1, x_1 \tan a) - x_2^2 (a + 2h \tan \beta + b \tan^2 \beta) = 0, \ldots\ldots\ldots\ldots\ldots(1)$$

and $\qquad x_2 [x_1 \{a + h (\tan a + \tan \beta) + b \tan a \tan \beta\} + (g + f \tan \beta)] = 0. \ \ldots\ldots\ldots(2)$

If $x_2 = 0$ the real points on the curve are obtained.

If $x_2 \neq 0$ the equation (2) is identically satisfied, if

$$g + f \tan \beta = 0 \quad \text{and} \quad a + h (\tan a + \tan \beta) + b \tan a \tan \beta = 0.$$

Hence the conic has an infinite number of points in one (a, β) plane of its nest, viz. the plane for which $\beta = \tan^{-1} \left(-\dfrac{g}{f} \right)$. This plane has its β parallel to the polar of the origin and its a is the line joining the origin to the centre of the conic. For no other plane of the nest are there points at a finite distance. For all planes outside the nest of the conic

$$x_1 = -\frac{g + f \tan \beta}{a + h (\tan a + \tan \beta) + b \tan a \tan \beta},$$

$$x_2^2 = \frac{f(x_1, x_1 \tan a)}{a + 2h \tan \beta + b \tan^2 \beta}.$$

Hence for given values of a and β, there is one value of x_1, and if the value of x_2^2 is positive, two values of x_2, which differ only in sign. Hence the corresponding points are a pair of conjugate imaginary points.

If the value of x_2^2 is negative there are no points on the branch considered. These results are geometrically consistent with those obtained in Art. 127.

(c) *An imaginary straight line.*

(1) *Origin the real point on the line.*

If the equation of an imaginary straight line is combined with the equation of its conjugate imaginary straight line an equation of the second degree is obtained, which does not involve the imaginary explicitly.

The equation may be graphed with respect to the real point on the lines, as if it represented a real conic.

Let the equation of the lines be $ax^2 + 2hxy + by^2 = 0$.

Branches are found to exist which correspond to values of a and β, which satisfy the relation

$$a + h\,(\tan a + \tan \beta) + b \tan a \tan \beta = 0.$$

That is, branches, which are graphically pairs of straight lines, exist in the planes for which a and β give lines, which are real pairs of harmonic conjugates of the pair of conjugate imaginary straight lines. These planes are said to form the nest of the straight lines. No points exist in other planes than those of the nest (Arts. 76 and 77).

(2) *Origin any real point.*

Let the equation of the line be

$$ax + by + c + i\,(a'x + b'y + c') = 0.$$

Then $\qquad x_1\,(a + b \tan a) - x_2\,(a' + b' \tan \beta) + c = 0,$ ....(1)

and $\qquad x_2\,(a + b \tan \beta) + x_1\,(a' + b' \tan a) + c' = 0.$ ....(2)

From (1) and (2)

$$x_1\,\{ac' - a'c + (bc' - cb')\tan a\} - x_2\,\{a'c' + ac + (b'c' + bc)\tan \beta\} = 0.$$

Hence, if $\tan a = \dfrac{ac' - a'c}{cb' - bc'}$ and $\tan \beta = -\dfrac{a'c' + ac}{b'c' + bc}$, there are an infinite number of points in the corresponding (a, β) plane.

Generally however there is only one point in an (a, β) plane, viz. that given by the equations (1) and (2). As a particular case there is only one real point on an imaginary line, i.e. the one point in the (1, 1) plane. If a line joins or contains two points in an (a, β) plane, that plane is termed the plane of the line, and it contains an infinite number of points in that plane. The relationship of these results to those given in (1) is obvious geometrically.

135. Table of graphic representation by means of (a, β) planes.

In the following, real and purely imaginary points, including the origin, are omitted.

Nature of locus	*Its equation*	*Planes in which there are points*
Real straight line.		
(1) Origin on line	$ax + by = 0$	The one plane $\left(\tan^{-1}\left(-\dfrac{a}{b}\right),\ \tan^{-1}\left(-\dfrac{a}{b}\right)\right),$ in which there are an infinite number of points.
(2) Origin any point	$ax + by + c = 0$	The planes $\left(a,\ \tan^{-1}\left(-\dfrac{a}{b}\right)\right)$, where a can have any value. There are an infinite number of points in each of these planes.

The Imaginary in Geometry

Nature of locus	*Its equation*	*Planes in which there are points*

Imaginary straight line—together with its conjugate imaginary line.

(1) Origin the real point on the line — $ax^2+2hxy+by^2=0$ — An infinite number of points in all (a,β) planes, for which a and β are harmonic conjugates of the lines. These planes form the nest of the lines.

(2) Origin any real point — $ax^2+2hxy+by^2+2gx+2fy+c=0$, where $\Delta=0$ — An infinite number of points in the (a,β) plane, the a axis of which passes through the real point on the line, the β being its harmonic conjugate with respect to the lines. One point on all other (a,β) planes except those of the nest, in which there is no point at a finite distance.

Real conic.

(1) Origin the centre — $ax^2+2hxy+by^2+c=0$ — An infinite number of points in all (a,β) planes, where a, β determine conjugate diameters, i.e. in the nest of the conic.

(2) Origin any real point — $ax^2+2hxy+by^2+2gx+2fy+c=0$ — An infinite number of points in the (a,β) plane, the a of which passes through the centre, the β being parallel to the conjugate diameter. Two points in all other (a,β) planes except those of the nest, in which there is no point at a finite distance.

Imaginary conic.

(1) Origin the centre — $S+iS'=0$ — Not more than four points in any (a,β) plane. If there are any points in an (a,β) or (β,a) plane, the sum of the points in these two planes is four (Art. 144).

(2) Origin any real point — $S+iS'=0$ — Same result as when the centre is the origin (Art. 144).

Imaginary conic—special case.

Origin the centre — $S+iS'=0$ — If $S'=0$ are a pair of lines which are conjugate diameters of $S=0$, there are an infinite number of points in the (a,β) plane, whose axes are the conjugate diameters $S'=0$ (Art. 145).

It follows that, as a straight line has always an infinite number of points in some (a,β) plane, the most general straight line can be found by joining two points in some (a,β) plane. A conic does not always contain an infinite number of points —or more than four points—in any (a,β) plane. Hence the most general form of a conic cannot be obtained by describing a conic through five points in the same (a,β) plane. The conic so obtained is the special case alluded to above.

136. Intersections of an imaginary line and a real conic.

An imaginary line must, by Art. 134, lie in one of the (a, β) planes of the centre of the conic.

(1) The line may lie in one of the planes forming the nest of the conic, in which case it has a continuous graph in that plane. This graph may or may not meet the graph of the conic in that plane. If it meets it, the line intersects the conic in two points in the (a, β) plane in question. If the two graphs do not intersect, the line meets the conic in two points in different (a, β) planes.

Let the equations of the line and conic referred to the axes of their common (a, β) plane be

$$\frac{x}{l} + \frac{iy}{m} = 1 \quad \text{and} \quad \frac{x^2}{a'^2} + \frac{y^2}{b'^2} = 1,$$

and let

$$K \equiv \frac{b'^2}{m^2} - \frac{a'^2}{l^2} \quad \text{and} \quad L \equiv 1 + K.$$

Then for one of the points of intersection

$$x = \frac{a'b'}{m} \frac{\sqrt{L}}{K} - \frac{a'^2}{lK}, \quad iy = -\frac{a'b'}{l} \frac{\sqrt{L}}{K} + \frac{b'^2}{mK}.$$

(a) If L is positive this point and also the other point of intersection, obtained by changing the sign of $\sqrt{L}$, are in the common (a, β) plane.

(b) If L is negative the coordinates of the point may be written as

$$x = -\frac{a'^2}{lK} + i\frac{a'b'}{mK}\sqrt{-L}, \quad y = -\frac{a'^2}{lK}\left(\frac{b'}{a'}\sqrt{-L}\right) + i\frac{a'b'}{mK}\sqrt{-L}\left(-\frac{b'}{a'}\frac{1}{\sqrt{-L}}\right).$$

If $y = Mx$, $y = M'x$ be the equations of the pair of lines which are the (a, β) axes of this point, $M = \frac{b'}{a'}\sqrt{-L}$ and $M' = -\frac{b'}{a'}\frac{1}{\sqrt{-L}}$, so that $MM' = -\frac{b'^2}{a'^2}$ and the lines in question are a pair of conjugate diameters of the ellipse.

The coordinates of the other point of intersection are obtained by changing the sign of $\sqrt{-L}$. Hence these two points are not in the same (a, β) plane.

(2) The line may lie in one of the planes which do not form part of the nest of the conic. It cannot then intersect the conic in two points in the same (a, β) plane, for it only contains one point in planes other than the plane in which it is situated. The two points of intersection may be constructed by means of Art. 49.

The tangent to the (a, β) branch of a curve at any point on that branch lies in the plane in question. Hence in the first of the cases dealt with in this article the pole of the line is in the (a, β) plane considered. Even in the second and third cases this is also true.

Intersections of straight lines and conics.

A real straight line intersects

(a) *A real straight line* in the $(1, 1)$ plane.

(b) *An imaginary straight line* in the (a, β) plane, in which β is parallel to the real line and a is the harmonic conjugate of β with respect to the imaginary line and its conjugate imaginary line.

(c) A real *conic* in the (1, 1) plane or in the (a, β) plane, in which β is parallel to the real line, and a is the conjugate diameter of this line with respect to the conic.

(d) An imaginary *conic* in two points in different (a, β) planes, which may be determined as follows. Obtain the equation of the pair of imaginary lines joining the centre of the conic—or any real point—to the points of intersection of the line with the conic. This can be done at once analytically. Then by (b) above the points of intersection of the real line with these lines can be obtained.

An imaginary straight line intersects

(e) A real *straight line* in the (a, β) plane, in which β is parallel to the real line and a is the harmonic conjugate of β with respect to the imaginary line and its conjugate imaginary line.

(f) An imaginary *straight line* in the (a, β) plane, in which β is parallel to one of two of the sides of the diagonal triangle of the quadrilateral determined by the imaginary lines and their conjugate imaginary lines. These sides are the axes of perspective of the two involution pencils, of which the imaginary straight lines are the double rays. The points of intersection of the imaginary lines are on these real axes of perspective and the corresponding a's are obtained by joining the real points on the imaginary lines to the mean points of the opposite vertices of the quadrilateral, which are situated on the axes of perspective.

(g) A real *conic* in two points generally in different (a, β) planes. The real lines on which these points are situated may be constructed as explained in Art. 49, and the corresponding values of the a's may be obtained as in (f) or from the equations of the line and conic (Art. 136).

(h) An imaginary *conic* in two imaginary points. These may be found as follows. Express the equations of the conic and straight line in the forms

$$\frac{x^2}{(a+ia')^2}+\frac{y^2}{(b+ib')^2}-1=0 \quad \text{and} \quad (A+iA')\,\frac{x}{a+ia'}+(B+iB')\,\frac{y}{b+ib'}-1=0.$$

Let the units of length along the axes be $a+ia'$ and $b+ib'$. Then the equations of the conic and straight line are respectively

$$x^2+y^2-1=0 \quad \text{and} \quad (Ax+By-1)+i\,(A'x+B'y)=0.$$

The points of intersection of these can be found by (g).

137. Properties of (a, β) figures.

Every real point is in the (1, 1) plane and every purely imaginary point is in the (i, i) plane. Every real point is also in all the (a, β) planes, the a axes of which pass through the point in question, and every purely imaginary point is in all the (a, β) planes, the β axes of which pass through the point in question. Thus all points are in the (a, β) planes, and the (1, 1) and (i, i) planes may be regarded as a regrouping of the real and purely imaginary points or as two planes of reference. From the latter point of view, given a definite origin, the planes (1, 1) and (i, i) may be regarded as superposed. Then if any line a be drawn in the (1, 1) plane through the origin and any line β be drawn in the (i, i) plane through the origin, the plane (a, β) is the plane determined by this pair of intersecting straight lines.

The following are some properties of (a, β) planes considered in relation to the $(1, 1)$ and (i, i) planes.

Every complex point may be defined by a real and a purely imaginary coordinate in some (a, β) plane. This construction is unique. The point in question is said to be situated in this (a, β) plane.

Through every complex point in an (a, β) plane an infinite number of straight lines may be drawn in that plane and one straight line in every other (a, β) plane, including the $(1,1)$ and (i,i) planes. [See (b) below.]

Every real point contains an infinite number of straight lines in the $(1, 1)$ plane and also an infinite number of straight lines in all the (a, β) planes, the a axes of which pass through the point in question. The lines in the $(1,1)$ plane are real and the lines in the (a, β) planes (with the exception of the axis) are imaginary.

A purely imaginary point has similar properties.

Every complex line may be defined by a real and a purely imaginary intercept on the axes of an (a, β) plane. This construction is unique. The line in question is said to be situated in this (a, β) plane.

On every complex straight line there are an infinite number of points in its (a, β) plane and in every other (a, β) plane, including the $(1,1)$ and (i,i) planes, there is one point on the straight line. [See Art. 134 (c) (2).]

Every real line contains an infinite number of points in the $(1, 1)$ plane and also an infinite number of points in all the (a, β) planes, the β axes of which are parallel to the given line. The points in the $(1, 1)$ plane are real and the points in the (a, β) planes (with the exception of points in which the axis of a meets the line) are imaginary.

A purely imaginary line has similar properties.

(a) *The line* $ax + by + c + i(a'x + b'y + c') = 0$ *lies in the* (a, β) *plane for which*

$$a = \tan^{-1} \frac{ca' - ac'}{bc' - b'c} \quad \text{and} \quad \beta = \tan^{-1} \frac{-ca - a'c'}{b'c' + bc}.$$

This follows from the fact that the line contains the real point and the purely imaginary point given by the equations

$$\frac{x}{\begin{vmatrix} b & c \\ b' & c' \end{vmatrix}} = \frac{y}{\begin{vmatrix} c & a \\ c' & a' \end{vmatrix}} = \frac{1}{\begin{vmatrix} a & b \\ a' & b' \end{vmatrix}}$$

and

$$\frac{ix}{\begin{vmatrix} b' & c \\ -b & c' \end{vmatrix}} = \frac{iy}{\begin{vmatrix} c & a' \\ c' & -a \end{vmatrix}} = \frac{1}{\begin{vmatrix} a' & b' \\ -a & -b \end{vmatrix}}$$

The connectors of these points to the origin give the a and β axes of the plane in which the line is situated.

The fact that every imaginary straight line is in an (a, β) plane for a given origin may also be proved as follows.

Let the equation of the line be

$$Ax + By + i(A'x + B'y + C') = 0 \quad \dots\dots\dots\dots\dots\dots(1)$$

referred to rectangular axes through the origin.

Let a line in the (a, β) plane meet the axes of that plane in points at distances a and ib from the origin. Referred to the axes of x and y the coordinates of these points are $a\cos a$, $a\sin a$ and $ib\cos\beta$, $ib\sin\beta$.

Hence the equation of the line joining these points is

$$x(a\sin a - ib\sin\beta) + y(-a\cos a + ib\cos\beta) + iab\sin(\beta - a) = 0. \ldots\ldots (2)$$

Comparing coefficients in (1) and (2), it follows that

$$\tan a = -\frac{A}{B}, \quad \tan\beta = -\frac{A'}{B'},$$

$$a = \frac{C'\sqrt{A^2+B^2}}{AB'-BA'}, \quad b = \frac{-C'\sqrt{A'^2+B'^2}}{AB'-BA'}.$$

Hence the equation of the imaginary line (1) can be uniquely expressed in the plane $\left(\tan^{-1}\left(-\frac{A}{B}\right), \tan^{-1}\left(-\frac{A'}{B'}\right)\right)$ in the form

$$\frac{X}{\sqrt{A^2+B^2}} + \frac{iY}{\sqrt{A'^2+B'^2}} = \frac{C'}{AB'-BA}$$

The axes of this plane are $Ax+By=0$ and $A'x+B'y=0$. The condition that the line (1) should lie in a plane of the nest of the conic $a'x^2 + b'y^2 + c' = 0$ is that

$$\frac{AA'}{BB'} = \tan a \tan\beta = -\frac{a'}{b'}.$$

(b) *To find the equation of the line through a given imaginary point which lies in a given (a, β) plane.*

Let the coordinates of the point be $x_1 + ix_2$, $x_1\tan a + ix_2\tan\beta$, and let the plane be the (a', β') plane.

The equation of the line may be written in the form

$$y - x_1\tan a - ix_2\tan\beta = (m + im')(x - x_1 - ix_2).$$

If this be the same line as

$$ax + by + c + i(a'x + b'y + c') = 0,$$

then by (a)

$$\tan a' = m - m'\frac{c}{c'} \quad \text{and} \quad \tan\beta' = m + m'\frac{c'}{c}.$$

Therefore

$$\frac{\tan a' - m}{-m'} = \frac{m'}{\tan\beta' - m} = \frac{c}{c'} = \frac{(\tan a' - \tan a)x_1}{(\tan\beta' - \tan\beta)x_2} = K \text{ (suppose)}.$$

Hence $$\frac{m}{K^2\tan\beta' + \tan a'} = \frac{m'}{-K(\tan a' - \tan\beta')} = \frac{1}{1+K^2},$$

and

$$m + im' = \frac{K\tan\beta' - i\tan a'}{K-i} = \frac{(\tan a' - \tan a)x_1\tan\beta' - i(\tan\beta' - \tan\beta)x_2\tan a'}{(\tan a' - \tan a)x_1 - i(\tan\beta' - \tan\beta)x_2}.$$

If the plane (a', β') is taken to coincide with the plane (a, β), then the only relation connecting m and m' is

$$(\tan a - m)(\tan\beta - m) = -m'^2.$$

This relation is satisfied by the $m + im'$ of every line in the plane (a, β).

(c) *To find the point on a given imaginary line, which is in a given (a, β) plane, any real point being the origin.*

Let OX and OY be the axes determining the $(a, β)$ plane, in which it is required to find the point on the given line. Let O' be the real point on the given line, and through O' draw $O'Y'$ parallel to OY. Let $O'X'$ be the harmonic conjugate of $O'Y'$ with respect to the line and its conjugate imaginary line. Let $O'P$ be the graph of the line for the axes $O'X'$ and $O'Y'$. Let OX meet $O'X'$ in N. Draw NP parallel to OY to meet the graph of the line in P. Then P is the required point. For its coordinates are the real length ON measured parallel to OX and the purely imaginary length NP measured parallel to OY.

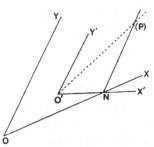

Hence it follows that the points on an imaginary line in the $(a, β)$ planes with any fixed real origin, obtained by keeping $β$ constant and varying a, are all on the graph of the line with its real point as origin for the plane, for which $β$ has the given value and a is the harmonic conjugate of $β$ with respect to the given line and its conjugate imaginary line.

138. The imaginary conic.

The general equation of an imaginary conic, which may be written shortly as $S + iS' = 0$, where $S = 0$ and $S' = 0$ are the general equations of two real conics, may be treated in either of two ways.

(1)   Transformations of the axes of coordinates may be limited to such as are real.

(2)   Transformations of coordinates may be allowed, which involve the imaginary.

In case (2), as a general rule certain points will be changed from real to imaginary and vice versâ. Transformations of type (1) will be considered in the first instance.

Transformations involving a real change of coordinates.

Such transformations depend on the fact that every conic has a real or semi-real inscribed quadrangle. One pair of opposite sides of such a quadrangle are real. Hence the equation of every conic can be obtained in the form

$$S + 2ih'xy = 0, \quad \ldots\ldots\ldots\ldots\ldots\ldots\ldots(1)$$

the axes of coordinates being a pair of real sides of the real or semi-real quadrangle which are inclined at an angle ω.

If the bisectors of the angles between the real sides be taken as the axes of coordinates, the coordinates are rectangular and the equation may be written

$$S + 2ih' (y^2 - m^2x^2) = 0. \dots\dots\dots\dots\dots(2)$$

If a, a', b, b' are the intercepts made by the conic on the pair of real sides of the quadrangle, (1) may be written in the form

$$bb'x^2 + aa'y^2 - bb' (a + a') x - aa' (b + b') y + aa'bb' + 2 (h + ih') xy = 0,$$
$$\dots\dots\dots(3)$$

where a, a', b, b' may be real, or purely imaginary and conjugate in pairs.

Generally the form of the equation of a conic depends on whether the inscribed quadrangle is real, semi-real of the first kind or semi-real of the second kind.

If the conic has a real self-conjugate triangle, the equation of the conic referred to this triangle as triangle of reference can be expressed in the form

$$ax^2 + by^2 + cz^2 + i (a'x^2 + b'y^2 + c'z^2) = 0. \dots\dots\dots(4)$$

139. Forms of the equation of a conic into which an imaginary conic can be projected by a real projection.

(A) (1) Let the conic have a real inscribed quadrangle.

Project the vertices into a real square, the diagonals of which are $2a$.

Take the diagonals for axes of coordinates.

The equation of any conic through the vertices is

$$x^2 + y^2 + 2hxy = a^2.$$

(2) Let the conic have a semi-real inscribed quadrangle of the first kind.

Project the vertices into a semi-real square of the first kind, the diagonals of which are $2ia$.

Take the diagonals as axes of coordinates.

The equation of any conic through the vertices is

$$x^2 + y^2 + 2hxy = - a^2.$$

(3) Let the conic have a semi-real inscribed quadrangle of the second kind.

Project the vertices into a semi-real square of the second kind, the diagonals of which are respectively $2a$ and $2ia$.

Take the diagonal of length $2a$ as axis of x and the other diagonal as axis of y.

The equation of any conic through the vertices is

$$x^2 - y^2 + 2hxy = a^2.$$

In the above h is or may be a complex quantity.

(B) (1) Let the diagonal triangle of the inscribed real or semi-real quadrangle be real.

Project one side to infinity and the other pair of sides into orthogonal lines.

Take these lines as axes of coordinates.

Then the equation of the conic is of the form

$$\frac{x^2}{(a+ia')^2}+\frac{y^2}{(b+ib')^2}=1 \quad \text{or} \quad (A+iA')\,x^2+(B+iB')\,y^2=1.$$

(2) Let the diagonal triangle of the inscribed semi-real quadrangle be semi-real. Project the pair of conjugate imaginary vertices into the critical points. Take rectangular axes through the centre of the conic, which is real. The equation of the conic is then of the form

$$\frac{(x+iy)^2}{(a+ia')^2}+\frac{(x-iy)^2}{(b+ib')^2}=1.$$

140. Transformation of axes involving the imaginary.

Let the equation of the conic be

$$(a+ia')\,x^2+2\,(h+ih')\,xy+(b+ib')\,y^2+2\,(g+ig')\,x+2\,(f+if')\,y+(c+ic')=0$$

or
$$a_1 x^2+2h_1 xy+b_1 y^2+2g_1 x+2f_1 y+c_1=0,$$

the axes being inclined at an angle ω.

If the origin be transferred to the point $\dfrac{F_1}{C_1}$, $\dfrac{G_1}{C_1}$ and the bisector of the angles between the axes of coordinates be displaced through an angle

$$\tfrac{1}{2}\tan^{-1}\frac{2h_1-(a_1+b_1)\cos\omega}{(a_1-b_1)\sin\omega},$$

the new axes being rectangular, the equation of the conic is

$$\frac{X'^2}{J_h+\sqrt{J_c}}+\frac{Y^2}{J_h-\sqrt{J_c}}+\frac{1}{J_p}=0, \quad \dots\dots\dots\dots\dots(1)$$

where
$$J_h=\tfrac{1}{2}\frac{C_1}{\Delta_1}\frac{a_1+b_1-2h_1\cos\omega}{\sin^2\omega},$$

$$J_p=\frac{C_1^2}{\Delta_1^2}\frac{C_1}{\sin^2\omega},$$

and
$$J_c=J_h^2-J_p=\tfrac{1}{4}\frac{C_1^2}{\Delta_1^2}\frac{(\overline{a_1+b_1}\cos\omega-2h_1)^2+(a_1-b_1)^2\sin^2\omega}{\sin^4\omega},$$

with the usual notation.

The equation (1) is of the form

$$\frac{X^2}{(a+ia')^2}+\frac{Y^2}{(b+ib')^2}=1,$$

or
$$(A+iA')\,X^2+(B+iB')\,Y^2=1.$$

141. Nature of an imaginary conic.

A conic may be imaginary either by nature or by displacement. Such a conic as $\dfrac{(x_1+ix_2)^2}{a^2}+\dfrac{y^2}{b^2}=1$, where x_2 is constant, is according to the definition of Art. 106 imaginary. If however the origin is moved through a distance ix_2 the equation becomes $\dfrac{x^2}{a^2} + \dfrac{y^2}{b^2} = 1$. Hence such a curve may be regarded as *imaginary by displacement*. If the constants which determine the nature of the curve, such as the semi-axes, the latus rectum or the angle between a pair of straight lines, involve " *i*," the curve is said to be *imaginary by nature*. This is the case when the quadratic equation for the axes of the conic involves the imaginary explicitly or when its roots are imaginary. In the latter case the equation of the curve is of the form

$$\frac{x^2}{(a+ib)^2} + \frac{y^2}{(a-ib)^2} = 1.$$

Thus with the notation of Art. 140, (1) if J_h and J_p are real and J_c is positive, the conic is imaginary by displacement, (2) if J_h and J_p are real and J_c is negative, the conic has a pair of conjugate imaginary axes, (3) if J_h or J_p is imaginary or complex, the conic generally is imaginary by nature.

142. Tracing of an imaginary conic.

The author is not acquainted with any satisfactory way of representing graphically the points on an imaginary conic in the general case. There are three possible ways of proceeding, but none of these seems to lead to satisfactory results.

(*a*) It is possible to transform the equation of the conic as explained in Art. 140 into the form

$$\frac{x^2}{(a+ia')^2} + \frac{y^2}{(b+ib')^2} = 1.$$

Lengths measured along the axis of x may then be regarded as multiples real or imaginary of $a+ia'$, and those measured along the axis of y as multiples real or imaginary of $b+ib'$. This method however does not seem to lead to satisfactory results.

(*b*) If as in the case of a real conic the points in (α, β) planes are sought, it is found that as a general rule (see Art. 143) there may be four such points in any plane, but generally not more than four points. Hence this method does not lend itself to graphic representation.

(*c*) It is possible to obtain a graphic representation by substituting for the systems of real parallel lines in an (α, β) figure systems of lines satisfying certain conditions, but the graphs obtained in this way are of a complicated nature.

Thus, if the equation of the conic be

$$ax^2 + by^2 + c + i\,(a'x^2 + b'y^2) = 0,$$

it is possible to construct the harmonic locus of $ax^2 + by^2 + c = 0$ and $a'x^2 + b'y^2 = 0$, which is a real curve. The tangents to this curve intersect the conic and its conjugate imaginary conic in conjugate imaginary points, the locus of whose mean points is

$$(ab' + ba')\{(ax^2 + by^2)^2 + (a'x^2 + b'y^2)^2\} + c\{y^2b^2a' + x^2a^2b'\}^2$$
$$- 4c\,(ab' - ba')^2 x^2 y^2\,(a'x^2 + b'y^2) = 0.$$

This curve would therefore intersect the tangents to the harmonic locus in points, which would give the real parts of the coordinates of the points on the conic, and a graph could be obtained by measuring off the appropriate imaginary lengths along the tangents. The same method could doubtless be employed, substituting any anharmonic locus for the harmonic locus.

The method (a) groups in the same figure points, whose coordinates referred to the same axes through the centre are of the form $k\,(a + ia')$, $l\,(b + ib')$, where k and l are real or purely imaginary quantities.

143. *In every* (a, β) *figure there may be four points, but not more than four points, on the conic*

$$ax^2 + 2hxy + by^2 + i\,(a'x^2 + 2h'xy + b'y^2) = c.$$

If $l + x_1 + i\,(l' + x_2)$ and $m + x_1 \tan a + i\,(m' + x_2 \tan \beta)$ be substituted for x and y and the real and imaginary parts of the resulting equation be equated to zero, there are no values of l, l', m, m', a, and β which render the two equations thus obtained identical or which make one of the equations identically zero in the general case.

If $l = l' = m = m' = 0$, the two equations are

$$x_1^2 T - 2x_1 x_2 C' - x_2^2 T_1 = c, \quad\dots\dots\dots\dots\dots\dots\dots(1)$$
$$x_1^2 T' + 2x_1 x_2 C - x_2^2 T_1' = 0, \quad\dots\dots\dots\dots\dots\dots\dots(2)$$

where $\quad C = a + h\,(\tan a + \tan \beta) + b \tan a \tan \beta, \quad T = a + 2h \tan a + b \tan^2 a,$

$\qquad\quad C' = a' + h'\,(\tan a + \tan \beta) + b' \tan a \tan \beta, \quad T' = a' + 2h' \tan a + b' \tan^2 a,$

$$T_1 = a + 2h \tan \beta + b \tan^2 \beta,$$
$$T_1' = a' + 2h' \tan \beta + b' \tan^2 \beta.$$

Looking upon (1) and (2) as the equations of two conics, the coordinates of a point on which are x_1, x_2, it is seen that for given values of a and β there are not more than four pairs of values of x_1 and x_2. Hence the result follows.

Equation (1) is that of a conic and (2) that of a pair of straight lines. Only real values of x_1 and x_2 are required. Hence for such to exist (2) must represent a pair of real lines and be equivalent to

$$T'\,(x_1 - \lambda_1 x_2)\,(x_1 - \lambda_2 x_2) = 0, \quad\dots\dots\dots\dots\dots\dots\dots(3)$$

where λ_1 and λ_2 are real.

From (1) and (2) $x_1^2(CT + C'T') - x_2^2(T_1C + T_1'C') = cC.$ ....(4)

Writing this equation as $x_1^2A - x_2^2B = cC,$ ....(5)

the values of x_2 are given by

$$x_2^2 = \frac{cC}{A\lambda_1^2 - B} \text{ or } \frac{cC}{A\lambda_2^2 - B}. \qquad\qquad\qquad (6)$$

The condition for the existence of points in a given (a, β) plane is that these expressions should be positive.

It may be noticed that if $x_1 + ix_2$, $x_1 \tan a + ix_2 \tan \beta$ be substituted for x and y in an equation,

(1) if ix_1 and ix_2 be values of x_1 and x_2, these values give a point $(-x_2, x_1)$ in the (β, a) plane;

(2) if a pair of values $\pm x_2$ correspond to a given value of x_1, the points so obtained are a pair of conjugate imaginary points;

(3) if $x_1 = \lambda x_2$, $\dfrac{y}{x} = \dfrac{\lambda \tan a + i \tan \beta}{\lambda + i}.$

Hence the point x, y lies on this imaginary line through the origin.

144.   *In the case of the conic*

$$ax^2 + 2hxy + by^2 + i(a'x^2 + 2h'xy + b'y^2) = c,$$

if there are any points in the (a, β) or (β, a) planes, the sum of the number of points in these planes is four.

Since there are points in one or other of the planes, equation (2) of the last article can be expressed as equation (3). Hence the values of x_2 are given by (6). These values are either real or purely imaginary. The real values give points in the (a, β) plane and the purely imaginary values give points in the (β, a) plane. The sum of the numbers of these points is four.

145.   Special case of an imaginary conic.

If in the case of the conic considered in the last article the lines $a'x^2 + 2h'xy + b'y^2 = 0$ are a pair of conjugate diameters of the conic $ax^2 + 2hxy + by^2 = c$, the relation (2), Art. 143, is satisfied, when a and β are such as to determine this pair of conjugate diameters. In this case $T' = T_1' = C = 0$. Hence there are in this case an infinite number of points on the conic in these (a, β) and (β, a) planes.

The corresponding values of x_1 and x_2 are given by

$$x_1^2 T - 2x_1x_2 C' - x_2^2 T_1 = c.$$

The conic in this case is not generally imaginary by displacement.

If a pair of real sides of the semi-real or real quadrangle of an imaginary conic are taken as the axes of coordinates and a new figure is obtained by projecting the side of the self-conjugate triangle opposite to their point of intersection into the line at infinity, it may happen that the vertices of the semi-real or real quadrangle of the imaginary conic so obtained are determined by a conic of which the axes of coordinates are a pair of conjugate diameters. In this case the equation of the imaginary conic is of the form

$$ax^2 + by^2 + 2ihxy = c.$$

Substituting $x_1 + iy_2 \tan\gamma$ and $x_1 \tan\alpha + iy_2$ for x and y, it is found that

$$a\,(x_1^2 - y_2^2 \tan^2\gamma) + b\,(x_1^2 \tan^2\alpha - y_2^2) - 2hx_1y_2\,(1 + \tan\alpha\tan\gamma) = c, \ldots(1)$$

and $\qquad x_1y_2\,(a\tan\gamma + b\tan\alpha) + h\,(x_1^2 \tan\alpha - y_2^2 \tan\gamma) = 0. \ldots\ldots(2)$

If $\alpha = \gamma = 0$, (2) is satisfied and (1) becomes

$$ax_1^2 - by_2^2 - 2hx_1y_2 = c.$$

Hence there are an infinite number of points in the $(1, i)$ and $(i, 1)$ planes of the conic.

Conic through five points in the same (a, β) plane.

If a conic be described through five points in the same (a, β) plane its equation by expanding the determinant, obtained by eliminating the constants, is found to be

$$ax^2 + 2hx\frac{y}{i} + b\left(\frac{y}{i}\right)^2 + 2gx + 2f\frac{y}{i} + c = 0, \qquad\ldots\ldots\ldots\ldots(1)$$

or $\qquad ax^2 - 2ihxy - by^2 + 2gx - 2ify + c = 0, \qquad\ldots\ldots\ldots\ldots(2)$

where a, b, c, f, g, h are real.

Let Δ_i be the discriminant of this equation, and Δ the discriminant of (1) when the "i's" are omitted. Let C_i and C be the corresponding minors of c.

Then, it will be found that

$$\Delta_i = -\Delta \quad \text{and} \quad C_i = -C.$$

Hence the equation referred to parallel axes through the centre is

$$ax^2 - by^2 + \frac{\Delta}{C} - 2ihxy = 0.$$

The axes $x = 0$, $y = 0$ are conjugate diameters of the curve $ax^2 - by^2 + \frac{\Delta}{C} = 0$, and therefore the conic considered comes under the special case of the imaginary conic.

Conic imaginary by displacement.

Consider the conic

$$ax^2 + 2ihxy - by^2 + 2gx - 2ify + c = 0,$$

where the axes are rectangular.

Referred to parallel axes of coordinates through the centre, the equation with the notation of the preceding is

$$ax^2 + 2ihxy - by^2 + \frac{\Delta}{C} = 0.$$

Turn the axes of coordinates through the imaginary angle $\frac{1}{2}\tan^{-1}\frac{2ih}{a-b}$.

The equation is then $a'^2x^2 + b'^2y^2 + \dfrac{\Delta}{C} = 0$, where a' and b' are given by

$$a + b = a' + b' \quad \text{and} \quad ab + h^2 = a'b'.$$

Therefore $\qquad\qquad a' - b' = \pm\sqrt{(a-b)^2 - 4h^2}.$

If $(a-b)^2 > 4h^2$, the values of a' and b' are real and the conic is imaginary by displacement. If this condition is not satisfied the axes of the conic are conjugate imaginary quantities.

In form (A), Art. 139, in cases (1) and (2), if h is purely imaginary the conic has a pair of conjugate imaginary axes. The same holds in case (3) except that when h in $i.h$ is <1, the axes of the conic are real and the conic is imaginary by displacement.

146. Modulus of reduction in an (a, β) figure.

In an (a, β) figure let P represent a point and let PN be y, ON be x, and the angle PON be θ.

Let $\beta - a$ be ω. Then

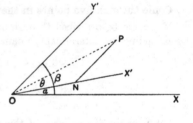

$$PN = x\,\frac{\sin\theta}{\sin(\omega - \theta)} = Kx \text{ (suppose)}.$$

Then

$$OP^2 \text{ graphically} = x^2 + K^2x^2 + 2x^2K\cos\omega,$$

$$OP^2 \text{ actually } = x^2 - K^2x^2 + 2ix^2K\cos\omega.$$

$$\therefore\ \frac{\text{actual value of } OP^2}{\text{graphical value of } OP^2} = \frac{1 - K^2 + 2iK\cos\omega}{1 + K^2 + 2K\cos\omega}$$

$$= R^2 \text{ (suppose)}.$$

$$\therefore\ R^2 = \frac{\sin^2(\omega - \theta) - \sin^2\theta + 2i\sin(\omega - \theta)\sin\theta\cos\omega}{\sin^2(\omega - \theta) - \sin^2\theta + 2\sin\theta\sin(\omega - \theta)\cos\omega}$$

$$= \frac{\sin\omega\sin(\omega - 2\theta) + 2i\sin(\omega - \theta)\sin\theta\cos\omega}{\sin^2\omega}$$

$$= \cos 2\theta - c.\sin 2\theta + 2i\sin\theta.c\,(\cos\theta - c\sin\theta), \text{ where } c \equiv \cot\omega,$$

$$= \cos 2\theta\,\{1 + ic^2\} + c\sin 2\theta\,\{i - 1\} - ic^2.$$

If $\qquad\qquad\qquad \omega = \dfrac{\pi}{2},\quad R^2 \Rightarrow \cos 2\theta.$

If $\qquad\qquad\qquad \theta = \dfrac{\omega}{2},\quad R^2 = \dfrac{i\cos\omega}{1 + \cos\omega}$

If $\qquad\qquad\qquad \theta = 0,\quad R^2 = 1.$

If $\qquad\qquad\qquad \theta = \omega,\quad R^2 = -1.$

Hence by means of R, the modulus of reduction, the actual distance from the origin of a point in an (a, β) figure can be deduced from its graphic representation.

The imaginary angle represented by a real graph.

Graphically a triangle ONP, in which ON is a, NP is ib, and ONP is a real angle ω, is represented by a real triangle ONP in which NP is b. Let the angle PON of the imaginary triangle be θ_i and that of the real triangle θ.

Then
$$\frac{b}{a} = \frac{\sin\theta}{\sin(\theta+\omega)} \quad \text{and} \quad \frac{ib}{a} = \frac{\sin\theta_i}{\sin(\theta_i+\omega)}.$$

Therefore
$$\frac{\sin\theta_i}{\sin(\theta_i+\omega)} = \frac{i\sin\theta}{\sin(\theta+\omega)} = iK \text{ (suppose)}. \quad\quad\dots(1)$$

Hence
$$\tan\theta_i = \frac{iK\sin\omega}{1-iK\cos\omega} \quad \text{and} \quad \tan2\theta_i = \frac{2iK\sin\omega\,(1-iK\cos\omega)}{1-2iK\cos\omega-K^2\cos2\omega} \dots(2)$$

From (1), if $\theta_i \equiv a+s\phi$, it is found by equating real and imaginary parts of the equation that
$$\tan2a = \frac{-K^2\sin2\omega}{1+K^2\cos2\omega} \quad \text{and} \quad \tan2s\phi = \frac{2iK\sin\omega}{1+K^2}. \quad\dots(3)$$

The value of $\tan2(a+s\phi)$ derived from (3) is in agreement with the value of $\tan2\theta_i$ derived from (2).

Given any imaginary length $A+iB$ measured along a line, it is possible to determine this length by means of (a, β) axes inclined at any given real angle ω.

Let a and ib be the coordinates of the end of the length $A+iB$.

Then
$$(A+iB)^2 = a^2 - b^2 + 2iab\cos\omega.$$

Assume that
$$A^2 - B^2 = a^2 - b^2 \quad \text{and} \quad 2iAB = 2iab\cos\omega.$$

Then
$$A^2 - B^2 = a^2 - b^2 \quad \text{and} \quad AB = ab\cos\omega.$$

$$\therefore\ a^2+b^2 = \pm\sqrt{(A^2-B^2)^2 + 4\frac{A^2B^2}{\cos^2\omega}} = \pm K, \text{ where } K \text{ is } > A^2-B^2.$$

$$\therefore\ 2a^2 = (A^2-B^2)\pm K \quad \text{and} \quad 2b^2 = -(A^2-B^2)\pm K.$$

If the positive sign of the square root is taken, a and b are both real. If the negative sign is taken for the square root both a and b are purely imaginary, that is, a is imaginary and ib is real.

147. (A, B) figures for the critical lines of a point.

Probably the two most remarkable of all the properties of the critical lines of a point are the following:

(1) If any pair of lines at right angles through the point be taken as axes of coordinates the equation of these lines is $x^2+y^2=0$. If real lengths be measured along the axis of x and imaginary lengths along the axis of y a graphic representation of the lines can be obtained. But if any other pair of lines at right angles through the point be taken as axes of coordinates the equation of the pair of critical lines is still $x^2+y^2=0$ and they can be graphed in the same way.

(2) Each of the critical lines may analytically be regarded as making the same angle with every real line in its plane (Art. 78).

The first of these properties can however be explained at once and the second is an immediate consequence of the first.

In Art. 76 it was shown that if the equation of a pair of imaginary lines is given in the form $y^2+m^2x^2=0$, i.e. if the axes of coordinates are the pair of real bisectors of the angles between the lines—which are of course at right angles—a graph exists in all the (a, β) planes for which a and β satisfy the relation $\tan a \tan\beta = -m^2$. If for the lines $y^2+m^2x^2=0$ be substituted the lines $y^2+x^2=0$, this condition becomes

$\tan a \tan \beta = -1$. Hence the axes for the graphs of these lines must be at right angles and for all pairs of axes at right angles a graph exists. Hence the property of the critical lines set forth in (1) is simply that for them as for other imaginary lines (a, β) figures can be constructed. For all pairs of axes at right angles the form of the equation of the critical lines is the same (Art. 96).

In the general case a pair of conjugate imaginary straight lines have only one pair of harmonic conjugates (real) which are at right angles. These are the bisectors of the angles between the lines. In Art. 77 it was shown that the graph of an imaginary line derived from its equation referred to two (a, β) axes makes with either of the bisectors the same angle as that deduced from its equation referred to the bisectors as axes of coordinates. In the case of the critical lines there are an infinite number of these bisectors. This result shows that the angle made by a critical line with any of these bisectors must be the same as that made by it with any one of another pair of bisectors, i.e. with any real line in the plane. Hence these two remarkable properties of the critical lines are only particular cases of well established theorems.

It is instructive to work out directly, by substituting $x_1 + ix_2$ and $x_1 \tan a + ix_2 \tan \beta$ for x and y, the (a, β) figures of a pair of critical lines (see Art. 76).

EXAMPLES

(1) Show that the line joining the points $x_1 + ix_2$, $y_1 + iy_2$ and $x_1 - ix_2$, $-y_1 + iy_2$ lies in the $(1, i)$ plane.

(2) Prove that the vertices of the real or semi-real quadrangle of the conic
$$ax^2 + 2hxy + by^2 + i(a'x^2 + 2h'xy + b'y^2) = c$$
lie in one or other of the planes $(1, 1)$, (i, i) or (a', β'), where a', β' give the directions of the common harmonic conjugates of $ax^2 + 2hxy + by^2 = 0$ and $a'x^2 + 2h'xy + b'y^2 = 0$.

(3) Prove with the notation of Art. 143 that the points of the conic, which lie in the (a, β) plane, are situated on the pair of straight lines whose equation is
$$T_1'(y - x \tan a)^2 + 2Ci(y - x \tan a)(y - x \tan \beta) + T''(y - x \tan \beta)^2 = 0.$$

(4) Prove that the graph of the curve
$$ax^2 + by^2 - 1 + i(by - lx)(ly + ax) = 0$$
in the (a, β) plane in which it has a continuous graph is the real part of the conic
$$\frac{bx^2}{l^2 + b^2} - \frac{ay^2}{l^2 + a^2} + \frac{xy(l^2 + ab)}{\sqrt{l^2 + a^2}\sqrt{l^2 + b^2}} = \frac{1}{l^2 + ab},$$
referred to the a, β axes as axes of coordinates.

(5) The locus of points, from which pairs of tangents can be drawn to a hyperbola such that the sum of the angles which they make with the axis of x is $\frac{\pi}{2}$, is a portion of a rectangular hyperbola.

(6) The locus of a point, the tangents from which to the ellipse $\frac{x^2}{a^2} + \frac{y^2}{b^2} - 1 = 0$ make angles with the axis of x the sum of which is $\frac{\pi}{2}$, is the rectangular hyperbola
$$x^2 - y^2 - a^2 + b^2 = 0.$$

(7) Prove that the point on the line $ax + iby = 1$ which is in the (a, β) plane is

$$\frac{a - ib \tan a}{a^2 + b^2 \tan a \tan \beta}, \qquad \frac{(a - ib \tan \beta) \tan a}{a^2 + b^2 \tan a \tan \beta}$$

(8) Show that if a, a', h, k, h', and k' are finite quantities and λ in the limit approaches the value zero, the connectors of the two points Ω_1 and Ω_2 whose co-ordinates are respectively $\dfrac{a}{\lambda} + h$, $\dfrac{ia}{\lambda} + k$ and $\dfrac{a'}{\lambda} + h'$, $\dfrac{-ia'}{\lambda} + k'$, to the origin are the critical lines of the origin.

(9) Hence show that the points Ω_1 and Ω_2 may be regarded as representing the circular points at infinity in the most general case and that these points in a particular case are a pair of conjugate imaginary points.

(10) Hence prove that in the general case the measures of the distances of the circular points at infinity from the origin are indeterminate quantities.

CHAPTER VII

THE IMAGINARY IN SPACE

148. In Chapter I an imaginary point was defined as a double point of a real overlapping involution situated on a real straight line. The real lines, on which such involutions were situated, were assumed to lie in a plane, so that the bases of the involutions were intersecting straight lines. In this chapter this restriction is removed so that the points considered, both real and imaginary, are situated in space.

A straight line may be regarded as a fundamental conception or may be defined as *a locus such that one and only one straight line can be drawn to pass through two given points*, and *a plane* as *the surface generated by straight lines all of which intersect in pairs at points at a finite or infinite distance*. It follows that two straight lines cannot intersect in more than one point. If they so intersected, two straight lines could be drawn to join their two points of intersection. It is assumed that two intersecting straight lines uniquely determine a plane.

A plane is determined by any three points A, B, C, which are not collinear. Join B and C to A by straight lines. Then, if P and Q be any two points on these lines, the straight line PQ intersects AB and AC and therefore the three lines lie in the same plane.

It follows that *a point and a straight line determine a plane*, and that *if two points on a straight line lie in a plane, every point in the straight line lies in the plane.*

The locus of points common to two planes is a straight line.

Let A and B be two points common to two planes σ and σ'. Only one straight line can be drawn to join A and B. Every point on the line AB lies in the plane σ and also in the plane σ'. Hence the theorem follows, for if there were a common point of the planes σ and σ', which was not on the line AB, the planes σ and σ' would coincide in the plane determined by this point and the line AB.

A plane and a straight line—which does not lie in the plane—determine a point.

Let σ be the plane and l the line. Take A any point in the plane σ and construct the plane σ' through A and the line l. This plane intersects the plane σ in a line $\sigma\sigma'$. l being in this plane meets $\sigma\sigma'$ in one point. This point is the point of intersection of l and σ.

Planes which intersect in the same straight line are termed coaxal planes.

Three planes which are not coaxal determine a point.

Let the planes be σ, σ', σ''. Then σ, σ' intersect in a straight line l, and l meets σ'' in a point. This is the one and only point common to the three planes.

Hence the following correlative relations are obtained:

Three non-collinear points determine a plane.	Three non-coaxal planes determine a point.
Two points determine a straight line.	Two planes determine a straight line.
A point and a line determine a plane.	A plane and a line determine a point.
Two intersecting straight lines determine a point.	Two intersecting straight lines determine a plane.

If in the above the determining elements are real, the line, point or plane so determined is real.

Parallel planes are such as contain more than one system of lines of the one parallel to a system of parallel lines of the other. Parallel planes form a coaxal system of planes the axis of which is entirely at infinity.

Analytical.

Let P be any imaginary point in space. Let P' be its conjugate imaginary point. Then PP' is a real line, for P and P' are by definition the double points of a real involution on a real straight line. Let M be the mean point of P and P', which is real. Through M draw a real plane σ perpendicular to the real line PP'. Then PM is a purely imaginary quantity. The position of M in the plane σ is determined by two real coordinates. Hence the coordinates of P are of the form $a, b, +ic$, and those of P' of the form $a, b, -ic$. If the planes of the coordinate axes are changed the coordinates of P and P' are of the form

$$a + ia', \quad b + ib', \quad c + ic',$$

and

$$a - ia', \quad b - ib', \quad c - ic'.$$

It follows that if the equation of the plane, which contains three given imaginary points, is of the form
$$(l + il')\, x + (m + im')\, y + (n + in')\, z = k + ik',$$
that of the plane which contains the three conjugate imaginary points is of the form
$$(l - il')\, x + (m - im')\, y + (n - in')\, z = k - ik'.$$

These planes in the equations of which the sign of the imaginary is changed are termed *conjugate imaginary planes*.

A line is determined by the equations of two planes which pass through the line. These may be transformed into the shape
$$\frac{x - (a + ia')}{(k + ik') - (a + ia')} = \frac{y - (b + ib')}{(l + il') - (b + ib')} = \frac{z - (c + ic')}{(m + im') - (c + ic')}.$$

1. *Every imaginary point has a conjugate imaginary point.*

If the coordinates of the point are
$$a + ia', \quad b + ib', \quad c + ic',$$
the coordinates of the conjugate imaginary point are
$$a - ia', \quad b - ib', \quad c - ic'.$$

2. *Through every imaginary point there is one and only one real line, which is the connector of the point to its conjugate imaginary point. Through this line pass all real planes which contain the point.*

The equation of the line joining the point $a + ia', b + ib', c + ic'$ to its conjugate imaginary point is
$$\frac{x - a - ia'}{a - ia' - a - ia'} = \frac{y - b - ib'}{b - ib' - b - ib'}$$
$$= \frac{z - c - ic'}{c - ic' - c - ic'}$$
or $\quad \dfrac{x - a}{a'} = \dfrac{y - b}{b'} = \dfrac{z - c}{c'}.$

This line is obviously real.

Every imaginary plane has a conjugate imaginary plane.

If the equation of the plane is
$$ax + by + cz + d$$
$$+ i(a'x + b'y + c'z + d') = 0,$$
that of the conjugate imaginary plane is
$$ax + by + cz + d$$
$$- i(a'x + b'y + c'z + d') = 0.$$

In every imaginary plane there is one and only one real line, which is the line of intersection of the plane and its conjugate imaginary plane. This line is the locus of all real points in the plane.

The real points in the plane
$$ax + by + cz + d$$
$$+ i(a'x + b'y + c'z + d') = 0$$
are given by
$$ax + by + cz + d = 0$$
and $\quad a'x + b'y + c'z + d' = 0.$

These points lie on a real straight line, which is also the real line in the conjugate imaginary plane.

3. *The line joining two imaginary points is the conjugate imaginary line of the line joining their conjugate imaginary points.*

4. *Two conjugate imaginary lines, determined as the connectors of two pairs of conjugate imaginary points, meet at a real point, if the real lines through the two pairs of conjugate imaginary points are concurrent, i.e. if the two pairs of conjugate imaginary points lie in the same real plane; otherwise they do not meet at all.*

The condition that the line

$$\frac{x - a - ia'}{k + ik' - a - ia'} = \frac{y - b - ib'}{l + il' - b - ib'}$$
$$= \frac{z - c - ic'}{m + im' - c - ic'}$$

should intersect the line obtained by changing the sign of i, is

$$\begin{vmatrix} a' & b' & c' \\ k' & l' & m' \\ k - a & l - b & m - c \end{vmatrix} = 0.$$

This is the condition that the real lines through the imaginary points determining the lines should meet at a point (see 2 above).

If this condition is not satisfied there is no real point on the conjugate imaginary lines and they do not intersect.

5. *If two imaginary points are such that the real lines through the*

The line of intersection of two imaginary planes is the conjugate imaginary line of the line of intersection of their conjugate imaginary planes.

Two conjugate imaginary lines, determined as the lines of intersection of two pairs of conjugate imaginary planes, lie in a real plane, if the real lines in the two pairs of planes are concurrent, i.e. if the two pairs of conjugate imaginary planes meet in a real point; otherwise they do not lie in a plane at all.

The condition that the line
$$ax + by + cz + d$$
$$+ i(a'x + b'y + c'z + d') = 0$$
$$kx + ly + mz + n$$
$$+ i(k'x + l'y + m'z + n') = 0$$
should intersect the line obtained by changing the sign of i, is

$$\begin{vmatrix} a & b & c & d \\ k & l & m & n \\ a' & b' & c' & d' \\ k' & l' & m' & n' \end{vmatrix} = 0.$$

This is the condition that the real lines in the determining planes should meet at a point, i.e. that they should lie in a plane (see 2 above).

If this condition is not satisfied there is no real plane through the conjugate imaginary lines and they do not lie in a plane.

If two imaginary planes are such that the real lines in the planes

points are not concurrent, no real plane can be drawn through their connector.

This follows from 4.

149. *Systems of lines and planes through a real point.*

A real point P contains

(1) An infinite number of real planes, which intersect in an infinite number of real lines through the point P. These planes are determined by P and any other two real points.

(2) An infinite number of planes determined by P, a real point A and an imaginary point Q. These planes may be real, in which case the plane contains Q', the conjugate imaginary point of Q.

All those planes which pass through a real point A as well as through P intersect in the real line AP.

(3) An infinite number of planes determined by P, Q and R, where Q and R are imaginary points. If a particular plane contains the conjugate imaginary points of Q and R, it is real. Otherwise it is imaginary.

Planes of this system intersect in imaginary lines through P.

Systems of lines and planes through an imaginary point.

Through an imaginary point P there passes one real line, which is the connector of P to its conjugate imaginary point P'.

Through the line PP' pass an infinite number of real planes, all of which contain P and P'. These are the only real planes which contain P. Hence if a real plane passes through an imaginary point, it also passes through the conjugate imaginary point.

are not coplanar, no real point exists on their line of intersection.

This follows from 4.

Systems of lines and points in a real plane.

A real plane σ contains

(1) An infinite number of real points, the connectors of which are an infinite number of real lines in the plane σ. These points are determined by σ and any other two real planes.

(2) An infinite number of points determined by σ, a real plane σ' and an imaginary plane σ_1. These points may be real, in which case the point contains σ_1', the conjugate imaginary plane of σ_1.

All those points which are contained by σ' as well as by σ lie in the real straight line $\sigma\sigma'$.

(3) An infinite number of points determined by σ, σ_1 and σ_2, where σ_1 and σ_2 are imaginary planes. If a particular point contains the conjugate imaginary planes of σ_1 and σ_2, it is real. Otherwise it is imaginary.

Points of this system lie on imaginary lines in σ.

Systems of lines and points in an imaginary plane.

In an imaginary plane σ there is one real line, which is the line of intersection of the plane with its conjugate imaginary plane σ'.

On the line $\sigma\sigma'$ there are an infinite number of real points, which are in the planes σ and σ'. These are the only real points in the plane σ. Hence if a real point lies in an imaginary plane it is also in the conjugate imaginary plane.

Through the line PP' also pass an infinite number of imaginary planes, which are conjugate in pairs. All of these planes have the line PP' for their real line and therefore contain no other real line.

In all these planes an infinite number of imaginary lines can be drawn through the point P.

Through the point P and any other two points, real or imaginary, an infinite number of imaginary planes can be drawn. These imaginary planes intersect in an infinite number of imaginary lines through P.

On the line $\sigma\sigma'$ are also an infinite number of imaginary points, which are conjugate in pairs. All of these points have the line $\sigma\sigma'$ for their real line and therefore contain no other real line.

Through these points an infinite number of imaginary lines can be drawn in the plane σ.

By means of the plane σ and any other two planes, real or imaginary, an infinite number of imaginary points can be constructed. These imaginary points determine an infinite number of imaginary lines in the plane σ.

150. Projection from an imaginary centre.

Projection from an imaginary centre of points in one real plane into points in another real plane.

Let S be the centre of projection, S' its conjugate imaginary point, σ the first plane and σ' the second plane.

(1) *All real points in one plane are projected into imaginary points in the other plane with the exception of the points where the line SS' meets the planes and the real points on the line of intersection of the planes σ and σ'.* The points which correspond to these points are real.

Consider any real point P in the plane σ other than the special points mentioned above. SP is an imaginary line since it does not pass through S', the conjugate imaginary point of S. It contains one real point P and therefore cannot contain any other real point. Hence P', the point where SP meets the plane σ', is an imaginary point.

(2) *Every conic is projected into a conic.*

Take A, B, C, D and P any five points on a conic L in the plane σ. Let these points be projected into the points A', B', C', D' and P'. Then $(P . ABCD) = (P' . A'B'C'D')$. Hence as P moves along the conic L, the point P' will move along the corresponding conic L'.

(3) *A real conic can always be projected into a real conic.*

Consider the conic L in the plane σ. Take any real point O in the plane σ outside the conic L, i.e. so that real lines can be drawn through the point O which meet the conic in imaginary points. Draw any real line s through O outside the plane σ and take two conjugate imaginary points S and S' on this line. Take S for centre of projection. Through O draw any real straight line a in the plane σ to meet the conic L in conjugate imaginary points A_1 and A_2. Then the plane $SS A_1 A_2$ is real and SA_1 is an imaginary line in this plane. It therefore contains a real point A_3.

Similarly draw through O in the plane σ two other straight lines b and c. From these derive as above points B_1 and C_1 and on their connectors to S determine the real points B_3 and C_3.

Take the plane $A_3 B_3 C_3$ as the plane σ'.

By construction the three imaginary points A_1, B_1, C_1 on the conic L project into the real points A_3, B_3, C_3 in the plane σ', which must be on the conic L' into which the conic L is projected by (2). Also the conic L meets the line $\sigma\sigma$ in a pair of points real, coincident or conjugate imaginary. Through these points the conic L' passes. Hence the conic L' contains three real points and a pair of points either real or conjugate imaginary. Hence the conic L' is real.

Instead of seeking the projection, in which A_1, B_1, C_1 are projected

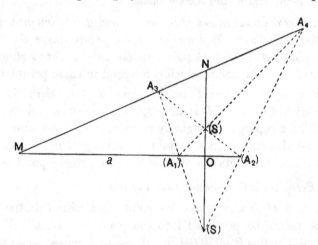

into A_3, B_3, C_3, it is possible to project A_1, A_2, B_1 into real points A_3, A_4, B_3, where A_4, B_4, C_4 are the points corresponding to A_2, B_2, C_2.

It should be noticed that A_3 is the point of intersection of SA_1 and $S'A_2$ and that the points B_3 and C_3 may be constructed in a similar manner.

Let SS' meet A_3A_4 in N and let A_3A_4 meet a in M. Since MOA_1A_2 is harmonic, M is a point in the polar of O with respect to the conic. Also $NOSS'$ is harmonic and therefore N is fixed when S, O, S' are fixed.

Consider the plane $A_3A_4B_3$. It meets σ in M a point on the polar of O. Also A_3A_4 and B_3B_4 intersect at N and therefore B_4 lies in this plane and is the projection of B_2 on the plane. Hence the plane $A_3A_4B_3B_4$ meets σ in the point K, where B_3B_4 meets b. Hence it intersects σ in the polar of O.

Similarly the plane $B_3B_4C_3C_4$ intersects σ in the polar of O. Hence since B_3, B_4 are common to the planes $A_3A_4B_3B_4$ and $B_3B_4C_3C_4$ these planes coincide. Hence (1) all pairs of conjugate imaginary points obtained by drawing real chords through O to meet the conic are projected into real points, and (2) for all positions of S the line $\sigma\sigma'$ is the polar of O.

(4) *All projective theorems, which are true for real points on a conic, are also true for imaginary points on a conic.*

Construct the projection as in (3) so that the real points for which the theorem holds are not the point SS'. σ or the points on $\sigma\sigma'$. Then the theorem is obvious.

Similarly it follows that projective theorems, which are true for real points on real straight lines, are true for imaginary points on real straight lines.

This affords a general confirmation of the results obtained in the earlier chapters.

151. Graphic representation of the imaginary in Solid Geometry.

Let the axes of coordinates be rectangular and let the coordinates of any point P be $x_1 + ix_2$, $y_1 + iy_2$, $z_1 + iz_2$.

Let P_1 be the point x_1, y_1, z_1, and let P_2 be the point x_2, y_2, z_2. Let O be the origin and let OP_1 and OP_2 be r_1 and r_2. Let the direction cosines of OP_1 and OP_2 be respectively $\cos\alpha$, $\cos\beta$, $\cos\gamma$ and $\cos\alpha'$, $\cos\beta'$, $\cos\gamma'$. Then the coordinates of P are $r_1\cos\alpha + ir_2\cos\alpha'$, $r_1\cos\beta + ir_2\cos\beta'$ and $r_1\cos\gamma + ir_2\cos\gamma'$. These values may be substituted in the equation of a surface in order to find the principal coordinates of the point P, which are r_1 and r_2, measured in the directions given by $\cos\alpha$, $\cos\beta$, $\cos\gamma$ and $\cos\alpha'$, $\cos\beta'$, $\cos\gamma'$.

(A)   *To trace a real plane.*

Substituting in the equation in its usual form it is found that

$$r_1 (l \cos \alpha + m \cos \beta + n \cos \gamma) - p = 0 \quad \dots\dots\dots(1)$$

and

$$r_2 (l \cos \alpha' + m \cos \beta' + n \cos \gamma') \quad = 0. \quad \dots\dots\dots(2)$$

From (2) it follows that the direction of the imaginary coordinate may be any direction parallel to the plane and that the coordinate r_2 may have any value. From (1) it follows that the point P_1 may be anywhere in the plane. Hence from the origin a real vector r_1 may be drawn to any real point in the plane and through that point an imaginary vector may be drawn in any direction in the plane and of any length to determine a point in the plane.

(B)   *To trace a real conicoid.*

Let the equation of the surface be

$$\frac{x^2}{a^2} + \frac{y^2}{b^2} + \frac{z^2}{c^2} - 1 = 0.$$

Substituting in this equation it is found that

$$r_1{}^2 \left\{ \frac{\cos^2 \alpha}{a^2} + \frac{\cos^2 \beta}{b^2} + \frac{\cos^2 \gamma}{c^2} \right\} - 1 - r_2{}^2 \left\{ \frac{\cos^2 \alpha'}{a^2} + \frac{\cos^2 \beta'}{b^2} + \frac{\cos^2 \gamma'}{c^2} \right\} = 0 \ \dots(1)$$

and

$$r_1 r_2 \left\{ \frac{\cos \alpha \cos \alpha'}{a^2} + \frac{\cos \beta \cos \beta'}{b^2} + \frac{\cos \gamma \cos \gamma'}{c^2} \right\} = 0. \ \dots(2)$$

Consider the line OP_1. It is a diameter of the conicoid whose direction cosines are $\cos \alpha$, $\cos \beta$, $\cos \gamma$. (2) is the condition that the line OP_2 should lie in the diametral plane of OP_1. If this condition is fulfilled (2) is satisfied.

Let the point P_1 be given. Then $P_1 P_2$ is parallel to the diametral plane of OP_1 and

$$r_2{}^2 = \frac{r_1{}^2 \left(\dfrac{\cos^2 \alpha}{a^2} + \dfrac{\cos^2 \beta}{b^2} + \dfrac{\cos^2 \gamma}{c^2} \right) - 1}{\dfrac{\cos^2 \alpha'}{a^2} + \dfrac{\cos^2 \beta'}{b^2} + \dfrac{\cos^2 \gamma'}{c^2}} \quad \dots\dots\dots(3)$$

Now

$$\frac{\cos^2 \alpha'}{a^2} + \frac{\cos^2 \beta'}{b^2} + \frac{\cos^2 \gamma'}{c^2} = \frac{1}{r^2},$$

where r is the length of the semi-diameter in which the line through the centre, whose direction cosines are $\cos \alpha'$, $\cos \beta'$, $\cos \gamma'$, meets the conic in which the diametral plane intersects the conicoid.

Therefore $r_2{}^2 = r^2 \left\{ r_1{}^2 \left(\dfrac{\cos^2 \alpha}{a^2} + \dfrac{\cos^2 \beta}{b^2} + \dfrac{\cos^2 \gamma}{c^2} \right) - 1 \right\} \dots\dots\dots(4)$

Hence the imaginary points on the conicoid which have P_1, as representing the real parts of their coordinates, lie in a plane through P_1 parallel to the diametral plane of OP_1 and at such distances from P_1 that they are situated on a conic similar and similarly placed to the section of the ellipsoid by the diametral plane of OP_1.

As the point P_1 moves along OP_1 the linear dimensions of the conic on which the points lie vary as

$$\sqrt{r_1^2 \left(\frac{\cos^2 \alpha}{a^2} + \frac{\cos^2 \beta}{b^2} + \frac{\cos^2 \gamma}{c^2} \right) - 1} \quad \text{or as} \quad \sqrt{\frac{x_1^2}{a^2} + \frac{y_1^2}{b^2} + \frac{z_1^2}{c^2} - 1}.$$

If OP_1 be taken as axis of z and the axes of x and y are any pair of conjugate diameters in the diametral plane of OP_1, and $\cos \alpha$, $\cos \alpha'$, ... are the direction ratios of r_1 and r_2, then

$$\cos \alpha = \cos \beta = \cos \gamma' = 0.$$

Also

$$r_1^2 \left(\frac{\cos^2 \alpha}{a^2} + \frac{\cos^2 \beta}{b^2} + \frac{\cos^2 \gamma}{c^2} \right) = \frac{z_1^2}{c^2}$$

and

$$\frac{1}{r^2} = \frac{\cos^2 \alpha'}{a^2} + \frac{\cos^2 \beta'}{b^2}$$

Hence (4) becomes

$$r_2^2 \left(\frac{\cos^2 \alpha'}{a^2} + \frac{\cos^2 \beta'}{b^2} \right) = \frac{z_1^2}{c^2} - 1.$$

Therefore

$$-\frac{x_2^2}{a^2} - \frac{y_2^2}{b^2} + \frac{z_1^2}{c^2} = 1.$$

This is the surface which is the graph of the branch in question of the conicoid.

This result may be obtained as follows. Take OP_1 as axis of z and two conjugate diameters in the diametral plane of OP_1 as axes of coordinates. Then, if x, y, z are the coordinates of points on the branch in question of the surface, z is real and x and y are imaginary. Hence x, y, z are respectively ix_2, iy_2 and z. Substituting in the equation of the surface it is found that the graph in question is $-\dfrac{x_2^2}{a^2} - \dfrac{y_2^2}{b^2} + \dfrac{z^2}{c^2} = 1$. This is an hyperboloid of two sheets which touches the real ellipsoid where it is met by the diameter OP_1. The same surface is obtained whatever pair of conjugate diameters in the diametral plane are taken for axes of x and y.

If OP_1 is taken as an imaginary axis, the coordinates of a point on the corresponding branch are x_1, y_1 and iz_2 and the equation of the graph is $\frac{x_1^2}{a^2} + \frac{y_1^2}{b^2} - \frac{z_2^2}{c^2} = 1$. This is an hyperboloid of one sheet, which touches the real ellipsoid where it is met by the plane $z = 0$. It is the same whatever pair of conjugate diameters in the plane $z = 0$ are taken as axes of coordinates. Hence for each diameter and its diametral plane, there are in addition to the $(1, 1, 1)$ and (i, i, i) branches, two hyperboloidal branches of the form $(i, i, 1)$ and $(1, 1, i)$. These two hyperboloidal branches have a common asymptotic cone $\frac{x^2}{a^2} + \frac{y^2}{b^2} - \frac{z^2}{c^2} = 0$.

Hence it is seen that the parent branch of the conicoid

$$\frac{x^2}{a^2} + \frac{y^2}{b^2} + \frac{z^2}{c^2} - 1 = 0$$

consists of eight parts, viz. the $(1, 1, 1)$ $(i, 1, 1)$ $(1, i, 1)$ $(1, 1, i)$ $(1, i, i)$ $(i, 1, i)$ $(i, i, 1)$ (i, i, i) branches. The branches 2 to 4 and 5 to 7 are of the same type.

The typical branches are

(a) $(1, 1, 1)$ branch which is the ellipsoid

$$\frac{x^2}{a^2} + \frac{y^2}{b^2} + \frac{z^2}{c^2} - 1 = 0.$$

(b) $(1, 1, i)$ branch which is an hyperboloid of one sheet

$$\frac{x^2}{a^2} + \frac{y^2}{b^2} - \frac{z^2}{c^2} - 1 = 0.$$

(c) $(i, i, 1)$ branch which is an hyperboloid of two sheets

$$-\frac{x^2}{a^2} - \frac{y^2}{b^2} + \frac{z^2}{c^2} - 1 = 0.$$

(d) (i, i, i) branch which does not exist in this case. Its equation is

$$\frac{x^2}{a^2} + \frac{y^2}{b^2} + \frac{z^2}{c^2} + 1 = 0.$$

The hyperboloids (b) and (c) have a common asymptotic cone

$$\frac{x^2}{a^2} + \frac{y^2}{b^2} - \frac{z^2}{c^2} = 0.$$

The conjugate loci may be obtained by taking any point A' on the ellipsoid and associating with OA' as axes of coordinates any pair of conjugate diameters OB', OC' in the diametral plane of OA'.

The equation of the ellipsoid is then

$$\frac{x^2}{a'^2} + \frac{y^2}{b'^2} + \frac{z^2}{c'^2} - 1 = 0.$$

As in the case of the parent branch there are eight branches of the curve, of which four are typical.

(a) (1, 1, 1) branch is the same ellipsoid as before.

(b) (1, 1, i) branch is an hyperboloid of one sheet which touches the ellipsoid where it is met by the plane $z = 0$. Its equation is

$$\frac{x^2}{a'^2} + \frac{y^2}{b'^2} - \frac{z^2}{c'^2} - 1 = 0.$$

(c) (i, i, 1) branch is an hyperboloid of two sheets, which touches the ellipsoid at the ends of the diameter, which is the axis of z. Its equation is

$$-\frac{x^2}{a'^2} - \frac{y^2}{b'^2} + \frac{z^2}{c^2} - 1 = 0.$$

(d) (i, i, i) branch does not exist in this case. Its equation is

$$\frac{x^2}{a'^2} + \frac{y^2}{b'^2} + \frac{z^2}{c'^2} + 1 = 0.$$

When it exists it is the same as the corresponding part of the parent branch.

The two hyperboloids (b) and (c) have the common asymptotic cone

$$\frac{x^2}{a'^2} + \frac{y^2}{b'^2} - \frac{z^2}{c'^2} = 0.$$

Consider the conicoid

$$\frac{x^2}{a^2} + \frac{y^2}{b^2} + \frac{z^2}{c^2} - 1 = 0$$

and any real plane parallel to the yz plane. Let the plane be $x - h = 0$ and let $h < a$. This plane meets the conicoid in the conic

$$\frac{y^2}{b^2} + \frac{z^2}{c^2} = 1 - \frac{h^2}{a^2}.$$

This is an ellipse. It has a real branch which corresponds to a real branch of the conicoid. It also has imaginary branches. Consider the branch of the ellipse for which y is real and z purely imaginary. This branch gives points on the conicoid of the form (1, 1, i). To obtain the principal coordinates of such points the x and y coordinates must be combined into a single real coordinate. This gives a real coordinate—say OP_1—in the xy plane and an imaginary coordinate perpendicular to this plane that is parallel to the axis of z. For this particular branch the coordinates of all points on the conicoid will be of this form. As however OP_1 will generally be different for different points, these points will not generally lie on the same branch of the surface.

The hyperboloids may be graphed in the same way as the ellipsoid. The branches will be found to be of the same nature. The paraboloids may also be graphed.

(C) *To trace an imaginary plane.*

Let the equation of the plane be

$$lx + my + nz + iK\,(l'x + m'y + n'z) = 0,$$

where $l^2 + m^2 + n^2 = 1$ and $l'^2 + m'^2 + n'^2 = 1.$

The equation of the conjugate imaginary plane is

$$lx + my + nz - iK\,(l'x + m'y + n'z) = 0$$

and the combined equation of the two planes

$$(lx + my + nz)^2 + K^2\,(l'x + m'y + n'z)^2 = 0.$$

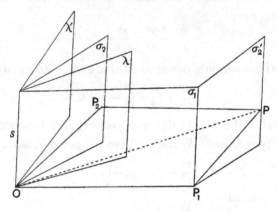

Let $lx + my + nz = 0$ be the plane λ, $l'x + m'y + n'z = 0$ the plane λ', and let s be the real line of intersection of the planes.

Let P_1 and P_2 be as previously defined. Then on substituting in the equation of the plane it is found that

$$r_1 . \Sigma l \cos \alpha - r_2 K . \Sigma l' \cos \alpha' = 0 \quad \ldots\ldots\ldots\ldots\ldots(1)$$

and $$r_2 . \Sigma l \cos \alpha' + r_1 K . \Sigma l' \cos \alpha = 0, \quad \ldots\ldots\ldots\ldots\ldots(2)$$

where $$\Sigma l \cos \alpha \equiv l \cos \alpha + m \cos \beta + n \cos \gamma,$$

$$\Sigma l' \cos \alpha = l' \cos \alpha + m' \cos \beta + n' \cos \gamma, \text{ etc.}$$

From (1) and (2)

$$\Sigma l \cos \alpha . \Sigma l \cos \alpha + K^2 . \Sigma l' \cos \alpha' . \Sigma l' \cos \alpha = 0. \quad\ldots\ldots\ldots(3)$$

This is the condition that the line OP_2 should lie in the diametral plane of OP_1 with respect to the pair of conjugate imaginary planes, that is, if the plane determined by P_1 and s be σ_1, and σ_2 be the plane harmonic conjugate of σ_1 with respect to the pair of conjugate imaginary planes, OP_2 must lie in the plane σ_2.

Wherever the point P_1 is situated in the plane σ_1, P_2 must lie in the plane σ_2; and conversely wherever P_2 lies in the plane σ_2, the point P_1 must lie in the plane σ_1. Hence the relation between the points P_1 and P_2 and the planes σ_1 and σ_2 is reciprocal.

Give P_1 a definite position in the plane σ_1. Then r_1, $\cos\alpha$, $\cos\beta$, $\cos\gamma$ are given. In the plane σ_2 draw a line OP_2 in any direction determined by $\cos\alpha'$, $\cos\beta'$, $\cos\gamma'$, so that relation (3) is satisfied.

Then from (1) $\qquad 1 = \dfrac{r_1 . \Sigma l \cos\alpha}{Kr_2 . \Sigma l' \cos\alpha'} = \dfrac{p_1}{K . p_2}$,

where p_1 is the perpendicular from P_1 on λ and p_2 is the perpendicular from P_2 on λ'. But p_1 is given and therefore p_2 is constant. Hence the locus of P_2 is a line (in the plane σ_2) parallel to s. If P_1 moves along a line parallel to s this line is unaltered. Hence as P_1 or P_2 moves in the planes σ_1 and σ_2 along a line parallel to s, the other point describes another line parallel to s. The distance $P_1 P$ equals OP_2, where P is in the plane σ_2', which can be drawn through P_1 parallel to σ_2. Hence the locus of P is a straight line obtained as the line of intersection of planes through P_1 and P_2 parallel to σ_2 and σ_1.

If P_1 moves along OP_1, i.e. if $\cos\alpha$, $\cos\beta$, $\cos\gamma$ are constant but r_1 varies, then the locus of P_2, for each value of r_1, is a straight line parallel to s. The same is true in regard to the locus of P.

Consider a section of the figure by a plane OP_1PP_2, where OP_1 varies but the directions of OP_1 and OP_2 are fixed.

Then from (1)

$$\frac{OP_1}{OP_2} = \frac{OP_1}{P_1P} = \frac{K\Sigma l' \cos\alpha'}{\Sigma l \cos\alpha} = \text{a constant.}$$

Hence, as P_1 moves along OP_1, the locus of P is a straight line OP through O. This is true for all planes OP_2PP_1. Hence the locus of P, when P_1 has any position in the plane σ_1, is a plane determined by one position of P and s the real line in the pair of conjugate imaginary planes. When both these conjugate imaginary planes are considered P may be in either of two such planes. In the preceding σ_1 and σ_2 may be any pair of planes harmonic conjugate of the given pair of conjugate imaginary planes. Hence the graph of a pair of conjugate imaginary planes consists of a system of pairs of planes which pass through their real line.

The coordinates of a point on such a graph may be constructed as follows. Let σ_1 and σ_2 be the (α, β) planes which are real harmonic conjugates of the plane and its conjugate imaginary plane. Take P_1 any point (real) in σ_1. Draw through P_1 a plane parallel to σ_2. It meets the graph in a line parallel to the real line in the imaginary planes. The coordinates of any point on this line are OP_1 real (where O is the origin) and P_1P imaginary.

Thus the coordinates of a point of the graph are any real vector in the plane σ_1 and an imaginary vector parallel to the plane σ_2.

The preceding may be illustrated as follows.

Consider two conjugate imaginary planes σ and σ', which intersect in a real line s. Draw a real plane λ perpendicular to s to meet s in O' and σ and σ' in a pair of conjugate imaginary lines l and l'. Consider the figure in the plane λ. In this plane there are an infinite number of real pairs of harmonic conjugates of l and l'. Let a and β be a pair of these lines. The graph of either of the lines l and l', when a and β are taken as axes, is a straight line in the plane λ. Let P be any point of this graph and let $O'P_1$ and P_1P (x_1 and iy_2) be its coordinates. Then the planes sl, sl', sa, $s\beta$ form a coaxal harmonic pencil. Referred to O any point on s as origin, the coordinates of P are real lengths OO' and x_1—which combine with a real length OP_1—and the imaginary coordinate iy. These are the principal coordinates of P. The locus of P is obviously one or other of two planes through s.

152. Solid perspective.

Consider any point S and any plane s (the centre and plane of perspective). Take any point in space A. Join S to A to meet s in N. Take a point A' on SA such that $(SNAA')=\lambda$ (a constant). Then A is termed the corresponding point of A in the perspective.

In this way if A be any point of a solid figure σ, then the point A' of another solid figure σ corresponds to A.

If A be a point on a straight line a, the point A' will be on another straight line a' which is said to correspond to a, and as the point A moves along the straight line a the point A' will move along the straight line a'. This follows from considering the plane perspective in the plane Sa. The corresponding lines a and a' intersect on the plane s.

If two lines a and b in the figure σ intersect, they determine a plane. Take any point P in this plane. Then any line l in this plane through P meets a and b. To P corresponds a point P'. To l corresponds a line l' through P' which intersects the lines a' and b' which correspond to a and b. Hence P' lies in the plane $a'b'$, which corresponds to the plane ab. Therefore to a plane corresponds a plane and corresponding planes intersect on s.

If any line SN be drawn through S to meet s in N and a point V be taken on SN such that

$$(SNV\infty)=\lambda,$$

then the locus of V is a plane parallel to s, since $\dfrac{SV}{NV}=\lambda$ (a constant). Also to every point on this plane there corresponds a point at an infinite distance in the second figure. But it has been shown that generally to a plane corresponds a plane. Hence it is assumed that all points at infinity in the second figure, and consequently in every solid figure, lie in a plane, which in this perspective corresponds to the plane in which V is situated, i.e. to the vanishing plane v.

The points at infinity in the first figure σ correspond to points in the second figure determined by

$$(SN\infty W)=\lambda$$

or by $\dfrac{NW}{SW}=\lambda$, $\therefore \dfrac{SV}{NV}\cdot\dfrac{SW}{NW}=1$. Hence the ratios of V and W are the same with respect to S and N, when these points are interchanged. Hence the distance of V from S is the same as the distance of W from N. These lengths may be measured on the perpendicular from S to s. Hence the locus of W is a plane w parallel to s and v.

Therefore if S be the centre, s the plane, and v and w the vanishing planes of a solid perspective,

(1) All points in the planes v and w correspond to points at an infinite distance in the other figure.

(2) If a plane a meets the vanishing plane v in the line av, then $S.av$ is parallel to a', and a' is the plane parallel to the plane (viz., $S.av$) drawn through the line $a.s$.

From this it follows that, from a certain point of view, all points at infinity may be regarded as lying in a plane. The method here sketched may be employed to find the properties of solid figures.

To prove that by a rotation round its real line, through an imaginary angle, an imaginary plane may be superposed on a real plane.

Every imaginary plane has an equation of the form

$$(ax+by+cz+d)+i\,(a'x+b'y+c'z+d')=0, \dots\dots\dots\dots\dots(1)$$

and the equation of its conjugate imaginary plane is

$$(ax+by+cz+d)-i\,(a'x+b'y+c'z+d')=0. \dots\dots\dots\dots\dots(2)$$

The real line s in these planes is the line of intersection of the real planes

$$ax+by+cz+d=0, \dots\dots\dots\dots\dots\dots\dots\dots\dots\dots\dots(3)$$

and $$a'x+b'y+c'z+d'=0. \dots\dots\dots\dots\dots\dots\dots\dots\dots\dots(4)$$

Through any real point A on this line draw a real plane σ perpendicular to this line. It will meet the planes (3) and (4) in real lines s_1 and s_2 and the conjugate imaginary planes (1) and (2) in a pair of conjugate imaginary lines s_1' and s_2'. In the real plane σ through the point of intersection of the conjugate imaginary lines s_1' and s_2' pass the pair of real lines s_1 and s_2, and as in Art. 51 the line s_1', by a rotation round A through an imaginary angle, may be brought into coincidence with the real line s_1. By this process the imaginary plane (1) is brought into coincidence with the real plane $ax+by+cz+d=0$.

Hence it follows that there is nothing essentially different between a real and an imaginary plane. As in the case of real and imaginary straight lines, the difference lies in their relation to points outside themselves.

EXAMPLES

(1) Show that in an imaginary projection all the real points on a real straight line except two can be projected into imaginary points on another real line.

(2) Prove that the real lines in the imaginary planes, which pass through an imaginary line, form a system of surfaces of the second degree.

(3) Prove that the real lines through the imaginary points, which lie on an imaginary straight line, form a system of surfaces of the second degree.

(4) Prove that an imaginary point $a+ia'$, $b+ib'$, $c+ic'$, contains a purely imaginary line the equations of which are

$$\frac{ix+a'}{a} = \frac{iy+b'}{b} = \frac{iz+c'}{c}.$$

(5) Prove that a purely imaginary plane contains a purely imaginary line which is parallel to the real line in the plane.

(6) In Art. 25 prove that if $S=S_1=S_2$, the values of all the anharmonic ratios of the four points are either $-\omega_1$ or $-\omega_2$, where ω_1 and ω_2 are the two imaginary cube roots of unity.

(7) Prove that if $(ABCD)=-\omega_1$, then $(ABCD)=(ACDB)=(ADBC)$.

INDEX OF THEOREMS

The reference numbers refer to articles

INDEX OF TERMS AND DEFINITIONS

The reference numbers refer to articles

CAMBRIDGE : PRINTED BY J. B. PEACE, M.A., AT THE UNIVERSITY PRESS

Printed in the United States
By Bookmasters